T0382670

Breeding and Culture of Freshwater Ornamental Fish

About the Author

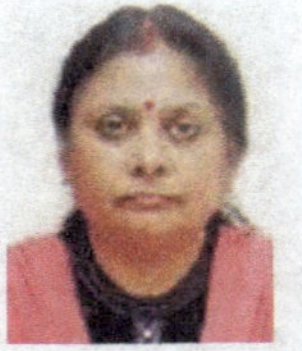

Dr. Archana Sinha, Principal Scientist, ICAR-Central Inland Fisheries Research Institute (ICAR-CIFRI), Barrackpore, Kolkata has working experience of more than 30 Years in the field of fisheries and aquaculture research, education and extension. She has completed several research projects and developed technology for fisheries and aquaculture, especially in ornamental fish breeding and culture. She was a course teacher for M.F.Sc., Ph.D. and Diploma course on ornamental fish. Guided student's for their Ph.D. thesis. She published more than 50 research papers on ornamental fish breeding and culture. She developed course module on ornamental fish breeding and culture for ICAR-CIFE (Deemed University), Mumbai; Skill Development Programme of ASCI, New Delhi. She is recognized as an expert for Ornamental fish breeding and culture by ICAR, New Delhi; NFDB, Hyderabad; MPEDA, Kochi; NABARD, Mumbai. She organized more than 100 short term training programmes on ornamental fish breeding and culture for unemployed, women, farmers and entrepreneurs for the development of ornamental fish culture. She has been conferred with several fellowships and institutional awards such as of Best Scientist award by Heeralal Chaudhuri Fisheries Foundation, Mumbai; NATCON Environmental Conservation Award by Nature Conservator, Muzaffarnagar; Gold Medal for outstanding research & academic contribution in the field of fisheries (ornamental fishes) by Zoological Society of India, Bodh Gaya.

About the Editor

Dr. Pramod Kumar Pandey has served for more than 32 years under National Agricultural Research System of India. Presently, he is serving as Dean, College of Fisheries, Central Agricultural University, Imphal. He has already served as Scientist, Senior Scientist and Principal Scientist in Indian Council of Agricultural Research before joining CAU, Imphal. He has more than 120 peer-reviewed international and national research papers, 3 books and several book chapters. Apart from that, he has delivered several invited talks and keynote addresses in numerous national and international meetings. He has been the Team Leader of the mega research project "Centre of Excellence in Fisheries and Aquaculture Biotechnology" funded by the Department of Biotechnology, Govt. of India. Since 2015, he is serving as President of the North East Society for Fisheries and Aquaculture, Agartala. Before that, he has served as General Secretary of Indian Fisheries Association for 5 years based at ICAR-CIFE, Mumbai, India. He has been conferred with several fellowships i.e. Zoological Society of India, Gaya; Society of Biological Sciences & Rural Development, Allahabad and national awards such as Best Teacher Award by HeeralalChaudhuri Fisheries Foundation, Mumbai; Paryayvaran Sanrakshak Award 2020 by Bharat TarunSangh, Alwar.

Breeding and Culture of Freshwater Ornamental Fish

Author

Dr. Archana Sinha

Principal Scientist

ICAR-CIFRI, Barrackpore, Kolkatta (W.B)

Editor

Dr. Pramod Kumar Pandey

Dean, College of Fisheries

Central Agricultural University

Lembuchera, Agartala (Tripura)

CRC Press is an imprint of the
Taylor & Francis Group, an **informa** business

NEW INDIA PUBLISHING AGENCY

New Delhi 110 034

First published 2024
by CRC Press
4 Park Square, Milton Park, Abingdon, Oxon, OX14 4RN

and by CRC Press
2385 NW Executive Center Drive, Suite 320, Boca Raton FL 33431

ISBN: 9781032599311 (hbk)
ISBN: 9781032599335 (pbk)
ISBN: 9781003456858 (ebk)

DOI: 10.4324/9781003456858

Typeset in Times New Roman
by NIPA, Delhi

Central Agricultural University
Lamphelpat, Imphal-795004 Manipur

Prof. S. Ayyappan
Chancellor

5 May, 2020

Foreword

Ornamental fish attracts human beings of all ages alike including children. Fish keeping has emerged as the second most popular hobby in recent years, next to photography. The sector's contribution is a very small but unique and vital part of an international fish trade. Considering great demand in the international market for indigenous fish of India and the large ornamental fish diversity in the country, it can be said that the sector has huge potential to earn foreign exchange and to create jobs for unemployed rural youth, especially women folk. The demand of individual hobbyists for such beautiful fishes has developed a global trade of ornamental fish involving the US $ 348 million and more than 60% of the production comes from the household of developing countries. The entire industry, including accessories and fish feed, is estimated to be worth more than the US $ 14 billion. However, the success of trade depends on the technology and human resource development (HRD) in the subject. As a hobby, people have developed a keen interest to know more and more about these fishes, their varieties, behaviour, and life span, breeding technology and rearing of the young ones. Therefore, the science behind the successful breeding and culture of ornamental fishes need to be explored.

I am delighted to note that the book entitled 'Breeding and culture of freshwater ornamental fish' is written by Dr (Ms) Archana Sinha, an author having a vast experience of teaching and research in the sector. I am sure that the readers will make use of the book for their skill development and to quench their curiosity. In addition to breeding and rearing technology of a number of important ornamental fishes, it also explains in detail the status of ornamental fish trade in the global market, diversity of ornamental fishes, their food and feeding habits, environmental issues, biosecurity, quarantine protocol, transportation and different Governmental schemes for the development of entrepreneurship among the rural youth and women folk. The illustrations and photographs add value to the book.

I am sure the book will be useful to the students and entrepreneurs alike, as also for planners to generate foreign exchange, human resource development and employment.

(S. Ayyappan)

Preface

It is well known that aquaculture is one of the fastest growing sectors in India. Even as growth in agriculture sector remains a challenge due to fluctuating growth in sectors like crop, livestock and forestry from 2014-15 to 2017-18, fisheries sector has grown rapidly from 4.9 per cent in 2012-13 to 11.9 per cent in 2017-18. Considering the fast growth of the sector, the target of fish production in India is fixed to be 20.00 million ton by 2022-23. Fish and fish product exports emerged as the largest group in agricultural exports and in value terms accounted for Rs. 47,620 crore in 2018-19. For economy of the country to grow to US$ 5.0 trillion by 2025, the fisheries sector has to play a very important role. In aquaculture, apart from food fishes, a substantial contribution has to come from ornamental fisheries, which has a great potential to increase the income of the farmers as well as in terms of earning foreign exchange.

The world trade of ornamental fish industry is of the tune of US$ million 348 but the contribution of India is only 1%, which is very meager, considering large number of ornamental fish species available in the country, having very attractive coloration, bands, pattern of color and demand in national as well as in international market. There are several indigenous fish species where an individual fish fetches several lakh rupees in the international market. That speaks volumes about the potentiality, scope and role of the sector. It can play as a growth engine of the economy. The bulk of the ornamental fish, involved in the trade, are collected from wild, raising the concerns about their sustainability and conservation in the nature. At the same time, small countries like Singapore have developed the breeding and rearing technology for many ornamental fishes and therefore, exporting it successfully to many countries of the world and capturing a lion share of the trade.

Keeping above points in the mind, this book has been written based on over three decades of experience of the author in the field of fisheries and aquaculture, especially in the field of ornamental fisheries. The book emphasizes on breeding, rearing, health management, water quality management and marketing strategy for ornamental fish. Apart from these it also addresses the issues of fish conservation, genetic improvements and related ancillary activities of the sector. The book is expected to help the small, marginal and large farmers alike in the development of ornamental fisheries. It will be also useful to the students' community, breeders, growers, traders and researchers as well.

Author

Contents

1

Status and Prospect of Ornamental Fish Culture

The culture of colored and attractive fish in glass aquariums for aesthetic use is called ornamental fish culture. The beautiful, tiny fishes or ornamental fishes are peaceful in nature and suitable for keeping in captivity. These lovely fishes are usually kept in a glass made aquarium and decorated with toys, plants, ceramic structures etc. for beautification. It displays the attractive fish that live in a natural environment decorated with accessories while maintaining environmental parameters in tanks/aquariums by using aerators, heaters, filters, lights to control water movement, temperature, suspended organic matter, illumination etc. besides feeding. Keeping ornamental fish in a glass tank is a very old and popular hobby. More and more people are getting attracted to this hobby and due to growing interest in aquarium keeping; it has resulted in a steady expansion in its trade in more than 125 countries. Household aquariums are more popular; therefore, less than 1% of the global market for ornamental fishes belongs to the public aquaria sector. Most of the ornamental fish is available from developing countries in the tropical and sub-tropical regions. The international trade in ornamental fish breeding and culture provides employment opportunities for thousands of rural people in developing countries. Ornamental fish is becoming an important component in Indian fisheries too, along with food fishes, for both income and employment generation. As a result of advancements in breeding, transport and rearing technology, more and more fish species are being recognized as ornamental fish almost every year. The ornamental fishery is recognized by many developing countries for employment generation and livelihoods. Ecologically suitable culture systems must be developed by evolving micro, small and medium enterprises for sustainable growth.

Evolution of ornamental fisheries

Archeological evidences of fish-keeping dates back to the Sumerians (2500 BC) and the Babylonians (500 BC). Egyptians considered fish sacred, worshiping the Nile Perch among others. Romans also kept fish in tanks but perhaps not for as decorative purposes as the Chinese; keeping them fresh for the dinner table. The Chinese kept carps and started breeding them selectively during the Sung Dynasty (960-1279). Records show that these fish were kept for purely decorative purposes; people were forbidden to eat them. Ornamental goldfish made its way into Europe by 1691. According to Tullock, the 17th century diarist, Samuel Pepys, referred to seeing fish being kept in a bowl and referred to the set up as "exceedingly fine".

First Sustainable Fish Tanks

While excited about the prospects of keeping fish indoors, fish enthusiasts did not understand how the water needed to be "cycled" in order for fish to stay alive for long indoors. In 1805, Robert Warrington is credited with studying the tank's requirement to be cycled to keep fish alive for longer.

Victorian England

With the opening of the public aquaria at the London Zoological Gardens at Regents Park in 1853, fish keeping as a hobby reached a new level of interest. In 1856, German Emil Robmaber wrote an essay, "Sea in a Glass," introducing fish keeping as a hobby to the public.This hobby required specialized equipment and attention at this point, reserving it for the wealthy. Fish tanks for tropical fish required heating via flames underneath (gas burning lamps underneath slate bottoms). When electricity was introduced into the home, fish enthusiasts began experimenting with electrical immersion heaters in glass tubes.

Commercial Fish Breeding

Until the 1920s, except for highly developed goldfish and carp keeping in Asia, most fish kept in tanks were captured from wild. In Florida in the 1920s, entrepreneurs began the first commercial fish breeding businesses.

America - Flying Fish

Until 1950s, most commercial fish breeders needed to situate themselves close to their demands. After World War-II, commercial fish breeders began to use ex-combat pilots to transport their fish around the world.

Better Tanks

In the 1960s, fish keeping as a hobby improved as the industry went from glass framed tanks to glass sealed tanks, allowing for better waterproofing of the tank. Further innovations include the advent of the acrylic tank, which is more lightweight, more crack resistant and lends itself to different shapes besides the basic rectangle glass tank.

Marine Tanks and Reef Keeping

The 1960s through the 1980s saw many developments in maintaining saltwater aquariums for the (albeit very serious) hobbyist as opposed to a more publicly funded zoo setting for saltwater tanks. Breakthroughs include understanding the role live rock plays in maintaining tank balances as well as advancements in filtration systems, including the use of protein skimmers and the wet-dry or trickle filtration methods. Understandings filtration systems, salinity need, and live rock requirements all helped to propel forward the saltwater tank for the home hobbyist.

History of ornamental fish keeping in India

Most of the freshwater ornamental fish belong to the family Cyprinidae, Balitoridae and Cobitidae under the order Cypriniformes. There are two hotspots of freshwater ornamental fish biodiversity in India, among which, the North-east India harbour about 250 species and the Western Ghats harbour about 155 species of indigenous ornamental origin. Two hundred and sixty one egg layers and 27 live-bearing exotic fish are very popular among the hobbyists in India. India offers a number of high priced freshwater ornamental fish like Barca snakehead, *Channa barca*, Kerala queen, *Puntius denisoni* etc.

According to Sane (1982), it was at Bombay in the first or second decade of the twentieth century that aquarium keeping commenced as a hobby on a small scale which led to the formation of societies in Madras and Bombay and especially the Taraporewala aquarium in 1951. The export started on an experimental basis in 1969 with foreign exchange earnings to the tune of US $ 0.04 million. Several aspects of ornamental fishes have been taken up of late. Premkumar and Balasubramanian (1984) studied the breeding biology of the Scarlet banded barb, *Puntius amphibius* from Chackai canal. Some researchers worked on the sexual dimorphism of freshwater puffer fish *Tetraodon travancoricus* (Hora and Nair), from Trichur District. Sunil et al. (1999) studied the length weight relationship in the Catfish *Horabagrus brachysoma* (Gunther). Sunil (2000) studied the length weight relationship in *Rasbora daniconius* (Ham) from Achenkoil River. Mercy et al. (2002) studied the length weight relationship of *Puntius denisonii*.

World Scenario

According to *FAO (2017)*, export earnings from ornamental fish trade was US $ 348 million and more than 60% of the production came from the household of developing countries. The wholesale value of the global ornamental fish trade is estimated to be US $ 1 billion while the retail value is US $ 6 billion. The entire industry, including accessories and fish feed, is estimated to be worth more than the US $14 billion.The top exporting country (with percentage contribution to global trade) is Singapore (19.8%), followed by Czech Republic (7.8%), Japan (7.4%), Malaysia (7.3%), Indonesia (5.3%), Israel (4.3%), Thailand (3.9%), Sri Lanka (2.9%) and India (0.008%). The largest importer of ornamental fish is the USA, followed by Europe and Japan. The emerging markets are China and South Africa.

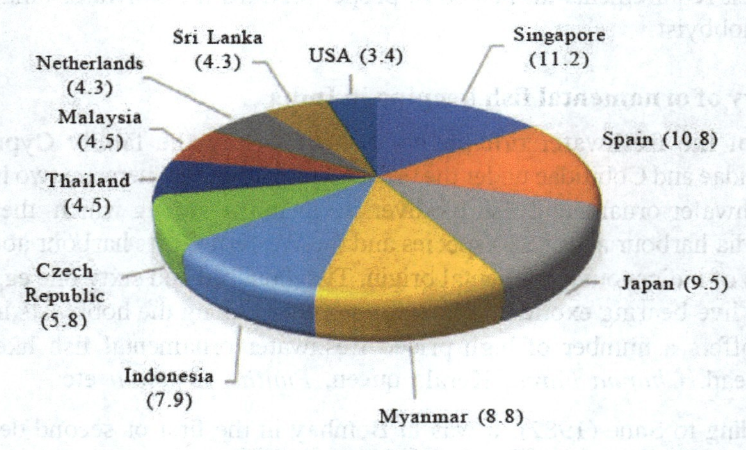

Fig. 1: Global ornamental fish export (2017)
Source : U.N. Data

More than 2,500 fish species are traded and some 30-35 species of freshwater fish dominate the market. The trade with an annual growth rate of 8 per cent offers a lot of scope for development. Individual hobbyists (home aquaria) control an overwhelming 99% of the market for ornamental fishes while only 1% of the market is controlled by public aquaria and research institutes. Global market demand is likely to grow to the tune of US $ 7 billion from the present level of US $ 5.26 billion. Singapore, being the largest producer of farm-bred ornamental fish handling (about 50% of the available species and varieties), is aptly called the "Ornamental Capital of the World". There are about 64 ornamental fish farms in Singapore that are registered – ten of these for the breeding of Dragon fish – occupying a total area of 133 ha. The Dragon fish or "Royal" fish that has a lifespan of 100 years is a protected species and can be traded only by permit; each fish could be fetching up to $ 50,000 in the retail market. Though Malaysia entered the field only 30 years ago, Penang is already famous for Discuss, Perak for Koi, Goldfish and Dwarf Gourami and Johore for live bearers like Guppy, Platy, Molly and Swordtail. Ornamental fish and aquatic plants have been assigned a priority in the Third National Agricultural Policy (1998-2010) of Malaysia with plans to produce 800 million ornamentals by 2010. In recent years, a mass propagation technique has been developed in Thailand to conserve the wild types of aquatic plants and is becoming an important industry. To promote the ornamental fish industry, the Thai government has set up an Ornamental Fish Research and Development Institute to provide training and technical knowledge to the local breeders to promote the export.

Indian scenario

India is lagging behind in ornamental fish trade and its overall domestic ornamental fish trade is worth about Rs. 555 crore and contribution to global export remain only 0.32%. Indian waters are considered as "JEWEL MINE" for domestic traders, exporters and hobbyists of ornamental fish. In India, the potential of ornamental fish is very high. As per an estimate of MPEDA, India has the potential to earn about US $ 5 billion as a foreign exchange by the export of ornamental fishes. Ornamental fish trade started in India in 1969 with export earnings of US $ 0.04 million. Indian ornamental fish sector is small and dominated by wild-caught species only. Ornamental fish activities are concentrated in 5 states i.e. West Bengal, Maharashtra, Karnataka, Tamil Nadu and Kerala. Indian exports mainly target South East Asia, China, Middle East, EU, USA and Japan. The major part of the export trade is based on wild collection from NEH. Out of 250 indigenous ornamental fish species, 155 fish species are being regularly exported and many fish species are also having very high potential.

An Overview of Indian Trade

Share of India to global ornamental fish export (*Source:* U.N. Data)

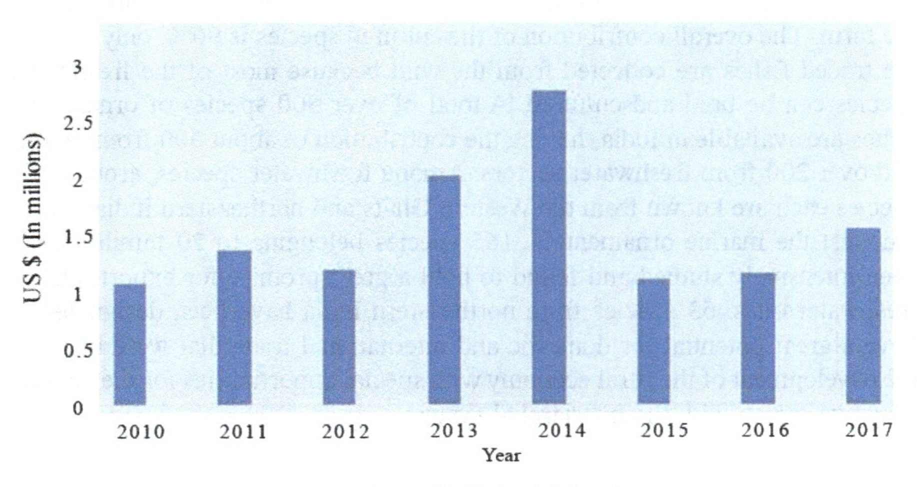

Fig. 2: Share of India in global trade

Kolkata dominates with respect to ornamental fish export trade, followed by Mumbai and Chennai. Registration as an exporter is granted under section 9(2) (h) of Marine Products Export Development Authority (MPEDA) Act 1972 read with rules 40-42 of MPEDA Rules, 1972. Registration is done for the following categories viz. Manufacturer, Exporter, Merchant, Route through Merchant & Ornamental Fish Exporter and also for entities such as Fishing Vessels, Processing Plants, Storage Premises, Conveyance, Pre-Processing Centers, Live Fish Handling Centre, Chilled Fish Handling Centre, Dried Fish Handling Centre, Independent Cold Storages and ice plants. There are total 25 registered exporters for ornamental fish, out of which the highest no. of 15 are from Kolkata and Chennai each, 11 from Kochi, 6 from Mumbai, 4 from Mangalore and 2 from Quilon (July 2014).

In addition to the export of ornamental fish for its wild-caught ornamental fish, the domestic market is also having very good demand, which is mainly based on captive-bred exotic species. About 80% of ornamental fishes are exported to the international market via Kolkata airport, of which major share comes from the North Eastern States of India. Other leading states in the trade are Kerala and Tamil Nadu. However, there is a vast unexplored potential for production of indigenous ornamental fishes and promoting ornamental fish culture in India. The scientific and systematic exploration of these potentials of ornamental fish will be a source of employment to women, Self Help Groups, entrepreneurs and unemployed youths to generate income, improve their livelihoods and earn considerable foreign exchange.The world's ornamental fish trade consists of about 80% freshwater species and 20% marine species whose contribution is increasing by establishing their breeding and rearing facility. Presently, 95% of marine fishes are collected from the wild and only 5% of fish are being bred in the farm. The overall contribution of the cultured species is 90%, only 10% of the traded fishes are collected from the wild because most of the freshwater species can be bred and cultured. A total of over 500 species of ornamental fishes are available in India, having the contribution of about 300 from marine and over 200 from freshwater sectors. Among freshwater species, around 100 species each are known from the Western Ghats and northeastern India, while, amongst the marine ornamentals, 165 species belonging to 20 families have been intensively studied and found to hold a great promise for export. Of the freshwater ones, 53 species from northeastern India have been designated to have a great potential for domestic and international trade that would help in the development of the rural economy with special opportunities for the gender-sensitive region with the matriarchal system.

Opportunities for women/unemployed youths

Women and youths have shown enthusiasm and expertise in different aspects of ornamental fish trade in India. They can be employed and gain in the following areas:

- Capture of fishes from the wild
- Culture of fishes
- Breeding of fishes
- Export of fishes
- Marketing of fishes and
- Marketing of accessories

The capture of fishes from wild

Wild ornamental fishes are abundant in those rivers and streams, which are flowing through dense forests and mountain terrains in India. These species such as devil catfish have good export potential and are ruling the foreign market of aquarium fish and are reaping a value of about US $ 1 to 2 per piece. In addition to these rivers and streams, the long coast line and several islands, which are stretching around with lagoons and coral reefs of India, abound in varieties of colourful marine fishes. These sources are presently exploited minimally but offer scope to enterprising persons to earn a livelihood. It is essential to create awareness among people, for them to take up the capture of these fishes and market them to earn maximum. Some of the indigenous fishes, which are often called trash fish, have been identified in the recent period as ornamental/aquarium fishes. The tiny colisa, loaches, danio, gourami of Indian origin are dominating in the market. However, no project has been undertaken by State Fisheries Departments on identification, survey, conservation, proper exploitation and mass production of ornamental fishes.

Culture of Ornamental fishes

For the culture of ornamental fish, the required infrastructure facilities have to be set up, supported by the application of relevant technical knowhow. Rearing of commercial ornamental species can be undertaken in re-circulation and flow-through water systems designed and established to maintain good water quality and to stimulate natural running water conditions. Different types of live feeds and artificial feeds are available in the market to rear ornamental fishes. Several workers pursue research work on the production of indigenous feed for these fishes. While in every major metropolitan city there are aquarists who own a few small ponds/cement tanks where they breed many freshwater ornamental fishes exclusively for domestic markets, this industry needs to be adequately popularized. Women aquarists are more caring for the small babies of tiny fishes. It is required to encourage them by providing technical know-how in local languages. Colourful handbooks on ornamental fish keeping and maintenance of aquariums are available for the hobbyists but the poor women entrepreneurs cannot afford that.

Breeding of ornamental fishes

The demand for ornamental fishes in domestic as well as international market is increasing rapidly. As such, sustainable exploitation of wild stocks of these fishes will not be able to meet the increasing demand. It is therefore essential to evolve appropriate breeding and rearing technology to produce both marine and freshwater ornamental fishes under controlled conditions in land-based infrastructural facilities. The technologies for breeding different varieties of ornamental fishes have now been established to such an extent that most of the aquarium fishes can not be bred as a household activity, both in rural and urban areas. Most of the aquarists breed only the common varieties of aquarium fishes like goldfish, guppies, platys, mollies, swordtails, gouramis, tetras, barbs etc., which are easy to breed. In order to enable householders to upgrade their capabilities, the State Government should come forward to encourage aquarists and interested entrepreneurs to take up farming of these highly-priced fishes. Simultaneously, technologies on the production of live fish food and nutritionally balanced dry feed in various forms such as pellets, powder, flakes, microcapsules etc., should be developed by technologists so that they can be extended to the hobbyists and entrepreneurs.

Export of ornamental fishes

In spite of having immense natural ornamental fish resources and technology for breeding and rearing them, not much of headway has been made in the country in the matter of export of ornamental fishes to foreign countries. So as to move ahead in these endeavours, MPEDA, Kochi has prepared a directory of ornamental fish exporters in which they have identified 25 ornamental fish exporters in India, especially, in Kolkata, Mumbai, Chennai and Kochi. The farmers and exporters have to be brought together, for the purpose of integrating the production and export activities in a manner that would be mutually beneficial. The establishment of such a relationship would push up the level of exports of ornamental fishes from the country, particularly to the USA, Europe and Japan. It has been reported that 8% of the estimated 86 million houses in the USA keep aquaria in their homes, 14% of the estimated 21 million houses in Great Britain, 4% of homes in Belgium and Holland and 5% of German and 20% of

Dutch houses keep fish. China, South Africa and several other countries too have the hobby of ornamental fish keeping. In view of the huge demand for export of ornamental fish, it is possible to undertake mass production of ornamental fish by farmers, to be made available to exporters. In fact, producers can become exporters so as to have the advantage of earning foreign exchange themselves. However, it is found that the women cooperative societies, which are breeding and rearing the ornamental fishes, are not getting justice in the market. They are not getting their reasonable share from exporters. It is due to lack of knowledge and communication problems. They are not aware of the export market and the outlets from where they can send the fishes directly. It is important to sensitize and make them aware about the marketing system.

Marketing of ornamental fishes

All the ornamental fishes, whether collected from wild, cultured or bred, have to be finally marketed. The women folk and/or unemployed youths, especially of rural India, can play a very important role in the marketing of the ornamental fishes. Like any other products, a marketing intermediary is the link in the supply chain that links the producer or other intermediaries to the end consumer. They are also known as middlemen or distribution intermediaries. The four types of marketing intermediaries are agents, distributors, wholesalers and retailers. The sector is capable of generating gainful employment to a large population and they, in turn, will help the sector to grow.

Fig. 3: Ornamental fish selling in weekly market in Kolkata

Marketing of accessories

In addition to the breeding, rearing and export of ornamental fishes, this trade has generated an ancillary business abroad. For beautification and maintenance of aquaria, rocks and gravels, artificial toys, natural and artificial plants, dry feed, live feed, aerators, filters are in use. There is a great demand for all these accessories. Different types of decorative toys with beautiful colorations, attractive shapes that are non-toxic to fishes are gaining popularity in the market. Submerged varieties of simulated aquatic ornamental plants from the natural habitat of ornamental fishes for placement in aquaria have a developing market. There are many aquatic plants for aquaria and some of them are costlier than ornamental fishes. Commonly available attractive aquatic plants are ribbon grass (*Vallisneria* spp.), arrow weed (*Sagittaria* spp.), spike rush (*Acorus* spp.), lace plant (*Aponogeton* spp.), faneard (*Cabomba* spp.), Indian water fern (*Ceratopteris* spp.), hornwort (*Ceratophyllum* spp.), Amazon Sword Plant (*Echinoderus* spp.), Hydrilla (*Hydrilla* spp.), Mint (*Ludwigia* spp.), Water Star (*Hygrophila* spp.), etc. Most of these plants can be grown and multiplied under controlled conditions. Artificial, non-toxic plants are also available in the market and are now increasingly attracting customers due to their blended colors and durability.

Apart from plants, a number of decorative toys are available for imparting an attractive look to an aquarium. They include plastic bubblers in the shape of a mermaid, underwater diver, oyster shell, angler human skull, tortoise, frog etc. There can be efforts to improve the materials, used for the manufacture of these toys, and also quality, texture, and color of these toys so that their utility can be enhanced, thereby providing a diversified activities status to the trade. Women can earn considerably from it even if they take it as a part-time engagement.

Fig. 4: Aquarium accessories

2

Ornamental Fish Diversity

India is one among the top ten mega-diverse countries of the world in terms of fish diversity (Dudgeon 2003). FishBase (Froese & Pauly 2016) has listed around 917 freshwater fish species (out of 2465 total fish species) as occurring in India. According to National Fisheries Development Board, Hyderabad, inland and marine waters in India possess a rich diversity of ornamental fish, with over 195 indigenous varieties reported from North-East Region and Western Ghats, and nearly 400 fish species from marine ecosystems. There are several other fish species which qualify to be called as ornamental fish. Considering enormous and diverse indigenous fish resources of the country, there is immense scope for India to become a potential candidate and a strong competitor in the international ornamental fish trade. Government of India has recognized ornamental fish sector as one of the thrust areas for generating employment opportunities and augmenting our foreign exchange earnings. It is estimated that the sector is directly and indirectly supporting about 50,000 of house-holds mainly in rural India. Large number of freshwater ornamental fish species is recorded from Indian waters and the majority of the global trade is based on freshwater exotic ornamental fishes including both classified and non-classified types.

Classified ornamental fishes

The small fishes like *Botia dario, Danio dangila, Puntius shalynius* and *Schisturareticulo fasciatus*, which can be reared in aquarium throughout their lifespan are called classified ornamental fishes.

Non-classified ornamental fishes

Larger food fishes like *Neolissochilus hexagonolepis, Labeo gonius, Channa marulius* and *Rita rita* which are treated as ornamental fish only in their juvenile stages are non-classified ornamental fishes.

Ornamental fish characteristics

Aquarium fishes are attractive due to their diversified ornamental values such as:

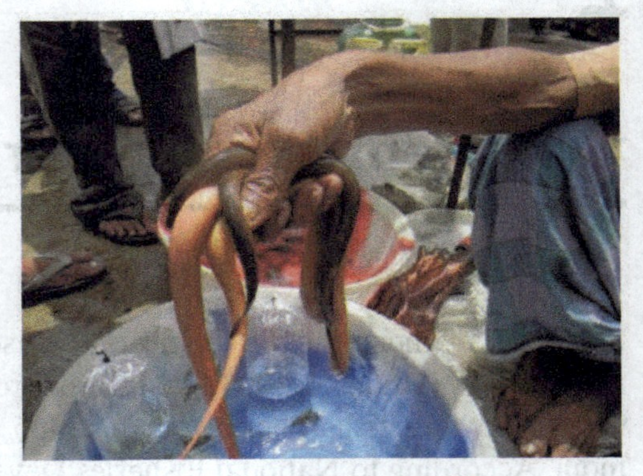

Fig. 5: Large ornamental fishes

- Beautiful colour (e.g. *Tetradon cutcutia, Colisa lalia*)
- Stripes and banding pattern (e.g. *Botia dario, Botia striata*)
- Attractive appearance (e.g. *Notopterus chitala*)
- Keeled abdomen (e.g. *Chela laubuca*)
- Peaceful nature and calm behaviour (e.g. *Ctenops nobilis*)
- Transparent body (e.g. *Parambassis baculis*)
- Hardiness (e.g. *Danio dangila, Brachydanio rerio*)
- Compatibility (e.g. *Puntius scheynius*)
- Beautiful jumping behaviour (e.g. *Esomus danricus*)
- Chameleonic habit (e.g. *Badis badis*)
- Charming predatory habit (e.g. *Glossogobius giuris*) and
- Longevity (e.g, *Anabas testudineus, Channa orientalis*)

Freshwater exotic ornamental fishes

There are large number of freshwater fishes which qualify as ornamental fish and the trade of which are flourishing in India. Common name of the important freshwater exotic ornamental fishes is given below:

Neon tetra (*Paracheirodon innesi*), Angel fish (*Pterophyllum scalare*), Siamese fighting fish (*Betta splendens*), Goldfish (*Carassius auratus*), Gourami (*Osphronemus goramy*), Discus (*Symphysodon aequifasciata*), Arowana (*Scleropages formosus*), Oscar (*Astronotus ocellatus*), Tiger Barb (*Puntigruste trazona*) and Danio (*Danio rerio*) are very popular in the ornamental fish trade. Except Arowana, all of these exotic ornamental fishes are commonly bred in India.

Marine ornamental fishes

Common clown (*Amphiprion percula*), False clown (*A. ocellaris*), Orange anemone fish (*A. sandaracinos*), three spot damsel (*Dascyllustri maculatus*),

Humbug damsel *(D. aruvanus)*, Blue damsel *(Pomacentrus caeruleus)*, Peacock damselfish *(P. pavo)* are very common and popular in India.

Classification of freshwater ornamental fishes

Freshwater ornamental fishes can be broadly classified into two groups. They are:

a) Live bearers &

b) Egg layers.

Fig. 6: Different types of ornamental fishes

Common live bearer fishes

Live bearer fish directly reproduce babies instead of spawning eggs. They are of following types.

1. Guppy (*Poecilia reticulata*)

Its origin is in South America, north of Amazon, but now it is enjoying worldwide distribution. It devours mosquito larvae thereby helping in the control of mosquitoes. These are tiny fishes with bright colors, looking very beautiful in groups. Male fish are more colorful than female and may reach up to 2.5 to 3.5 cm in length, while the female is usually larger in size when fully grown. They grow better in the water having temperature ranges 20-25 °C.

2. Swordtail (*Xiphophorus helleri*)

It has originated from Central and North-Eastern South America. The identifying character is the magnificent sword-like extension formed by the lower rays of the caudal fin in the male fish, which serves the purpose as an adornment. The fish prefers slightly saline water and voraciously devours live feed. The usual length of the female fish is 12 cm while that of the male is 8 cm. The phenomenon of sex reversal is observed in this species.

3. Platy (*Xiphophorus maculatus*)

Platys are of different types-namely red platy, orange platy, green platy and duckcido platy according to the colour. It also originated from Central and North-Eastern South America. The usual length of male platy ranges from 4- 4.5 cm and that of female from 5 to 5.5 cm. They breed once in three weeks and deliver about 75 young ones every time.

4. Mollies (*Poecilia reticulata*)

Origin of the fish is same as that of platy and swordtail. These fishes are easily bred and the usual length reaches 9-10 cm. They prefer saline water and breed every month, delivering around 250 young ones at a time. They reach marketable size within two months.

Common egg layer fishes

Most of the aquarium species are egg layers showing external fertilization. Fishes of this group can be further divided into five sub-groups.

1. Egg scatters

These species scatter their adhesive or non-adhesive eggs to the substrate viz. plants or allow them to float on the surface. The egg-scatters either spawn in pairs or in groups. There is no parental care and even they eat their own eggs. They produce a good number of eggs. *e.g.* goldfish (*Carassius auratus*) and tetras.

2. Egg depositors

In this case, the eggs are either laid on a substrate, like a stone or a plant leaf or even individually placed among fine-leaved plants like java moss. Egg depositors can be categorized into two groups- one that cares for their eggs and another that does not care. The egg depositors that care for their eggs are cichlids and some catfish. Cyprinids and various catfish come under the egg-depositors group that does not care for their young ones. These species lay their eggs on a surface and the eggs are abandoned. These species usually do not eat their eggs. *e.g.* Angel (*Pterophyllum scalare*), Discuss (*Symphysodon discuss*) and some catfish.

3. Egg bearers

Fishes in this group usually live in the seasonal waters that dry up at some time of the year. The majority of egg bearers lay their eggs in the mud. The maturity time for parents is very short and they lay eggs before dying when the water dries up. The eggs remain in a dormant stage until rains stimulate hatching. *e.g.* Annual killfish.

4. Mouth brooders

Mouth-brooders, as the name suggests, carry their eggs or larvae in their mouth. Mouth brooders can be grouped into ovophiles and larvophiles. Ovophiles are egg loving mouth brooders. They lay their eggs in pits and then the female sucks up into the mouth. The small numbers of large eggs hatch in the mother's mouth and the fry remains there for some time. Many cichlids and some labyrinth fish are the examples of ovophiles mouth-brooders. In contrast, larvophile or larvae loving mouth-brooder fish lay their eggs on a substrate and guard them until the eggs hatch. After hatching the female picks up the fry and keeps them in her mouth. When the fry can feed for themselves, they are released. e.g. Cichlids.

5. Nest builders

There are some fish species that build different types of nests for their eggs. The nest ranges from a simple pit, dug into the gravel to the elaborate bubble nest formed with saliva-coated bubbles e.g. Gouramis, Anabantids and some catfish.

Guppy Sword tail

Black molly Platy

Fig. 7: Live bearer ornamental fishes

Angel fish Goldfish

Koi carp
Tiger barb
Siamese fighter
Blue gourami
Pearl gourami
Devil catfish
Discuss
Oscar

Fig. 8: Common freshwater ornamental fishes

Fig. 9: Marine ornamental fishes

Commercially important ornamental fish

The top ten groups of ornamental fishes are the tetra, guppy, goldfish, catfish, molly, gourami, platy, loach, cichlid and the barb. Of the 30-35 favourite species of aquarists, only a few are Asian in origin. The most common are *Brachydanio rerio* and *Puntius conchonius*. Now-a-days preference is given for keeping large-sized fishes in the aquaria, probably due to their hardy nature and attracting visibility. The National Bureau of Fish Genetic Resources (NBFGR), Lucknow organized a workshop at Cochin in 1998 in which about 30 species of highly-priced ornamental fishes were identified for culture including *Puntius denisoni*. The hillstream fishes belonging to the genera *Balitora*, *Barilius*, *Garra*,

Homaloptera, *Lepidocephalus*, *Nemacheilus* and *Psilorhynchus* are considered to be coldwater ornamental fish species. These are found in warmer waters too and could be easily acclimated to the stagnant water conditions found in the aquaria. Some of the other endemic species from the south are *Aplocheilus lineatus*, *A. blockii*, *Danio malabaricus*, *D. aequipinnatus*, *Macropodus cupanus*, *Oryzias melastigma*, *Pristolepis-marginata*, *Puntius melanampyx*, *P. mahecola*, *P. arulius*, *P. narayani*, *P. setnai*, *Etroplus maculatus* and *E. canarensis* that are known to have an immense potential for export.

Table 1. Commercially important fishes in global market

Scientific name	Common name	Distribution	Remarks
Family-Cyprinadae			
Barbus everetti	Clown barb	Singapore, Indonesia, Malaysia	Rivers, ponds; 14 cm; males colorful
Barbus tetrazona	Tiger barb	Indonesia, Malaysia	Running waters; 7 cm; males colorful
Brachydanio rerio	Zebra barb	Eastern India and Bangladesh	Sluggish waters; 6 cm; males colorful
Carassius auratus	Gold fish	China, now everywhere	Over 100 colourful varieties; 30 cm/15 years
Danio aequipinnatus	Giant Danio	Sri Lanka and Eastern India	Sluggish standing waters; 10 cm; females larger
Puntius conchonius	Rosy barb	India and Bangladesh	Rivers and ponds; 14 cm; males attractive;
Family-Anabantidae			
Betta splendens	Siamese fighting fish	Thailand, Kampuchea, Vietnam	Males larger (6 cm), colorful; carnivorous
Colisa chuna	Honey gourami	India, Myanmar, Thailand, Malaysia	Males colorful; 5 cm; omnivorous
Colisa lalia	Dwarf gourami	India and Bangladesh	Males larger (5 cm), brightly colored;
Helostoma-temminckii	Kissing gourami	Malaysia and Thailand	Males fight in a kissing posture; 30 cm;
Trichogaster leeri	Pearl gourami	Malaysia, Thailand and Indonesia	Males colorful; 10 cm; omnivore
Trichogaster-trichopterus	Blue gorami	Malaysia, Thailand, Myanmar	15 cm; males build bubble nests for egg-laying

Family-Characidae			
Hemigrammu-serythrozonus	Glow light tetra	Brazil, swamps and rivulets	Females larger; 4 cm;
Hyphessobryconery-throstigma	Bleeding heart tetra	Brazil, Colombia and Peru	Males brightly colored; 12 cm
Metynnis hypsauchen	Silver dollar	Brazil	Several varieties; 10 cm
Paracheirodon innesi	Neon tetra	Brazil	Females larger; 4 cm; community fish
Family-Poecilidae **(Live bearers)**			
Poecilialatipinna	Sailfin molly	Central America to Mexico	Females larger (18 cm); do well in salt water too
Poecilia reticulata	Guppy	Central America to Mexico	Females larger (8 cm); variety of colors
Xiphophorus helleri	Swordtail	Southern Mexico and Gautemala	Females larger (12 cm); male with swordtail
Xiphophorus maculatus	Platy	Yucatan Peninsula and Mexico	Females larger (6 cm); no sword, dorsal rounded
Family-Cichlidae			
Cichlasoma meeki	Firemouth cichlid	Central America, Mexico, Yucatan	Males colorful, longer dorsal; 12 cm
Pterophyllum scalare	Angel fish	South America	Disc-like (12 cm long/ 25 cm deep); varieties
Symphysodonae-quifasciatus	Discus	South America	As deep as long (20 cm); sexing difficult
Family-Loricariidae			
Hypostomus-plecostomus	Pleco	Brazil and Peru	Need shelters, high oxygenation & filtration
Xenocaradoli-chopterus	Suckermouth	South America	15 cm; other species up to 60 cm
Family-Callichthyidae			
Corydoras aeneus	Bronze cory	South America	Long courtship before spawning; 7.5 cm
Family-Cobitidae			
Acanthoph-thalmuskuhlii	Kuhli loach	Southeast Asia	Prefer fine sandy substrate and dense vegetation
Botia macracanthus	Clown loach	Indonesia and Malaysia	42 cm;make audible sounds and have eye spines
Family-Belontiidae			
Macropodus opercularis	Paradise fish	Asia	10 cm; Bubble nest builders; male guards

A report shows that a few of the indigenous freshwater fish has also been bred successfully. They are *Colisa sota*, *C. fasciata*, *Oreichthys cosuatis*, *Gagata cenia*, *Danio dangila*, *Nandus nandus*, *Puntius melanampyx*, *Puntius melanostigma*, *Puntius filamentosus*, *P. vittatus*, *Parluciosoma daniconius*, *Pristolepis marginata*, *Garra mullya*, *Nemacheilus triangularis*, *Danio malabaricus*, *Esomus danricus*, *Etroplus maculatus* and *Macropodus cupanus*. Of the marine species, the clownfish and sea horses are of considerable importance.

Danio | Loach | Freshwater shark | Badis | Glass fish | Eel | Dwarf gourami | Rosy barb

Fig. 10: Indigenous ornamental fishes

North eastern states as hotspots of ornamental fish

India's northeastern states have very unique natural and geographical features ranging from the high-altitude terrain of the eastern Himalayas to the Brahmaputra flood plains and rain forests to vast wetlands. The region is very distinct in having certain endemic genera of fishes, viz., *Aborichthys, Akysis, Badis, Bangana, Chaca, Conta, Erethistoides, Erethistis, Exostoma, Myerglanis, Olyra, Parachiloglanis, Pareuchiloglanis, Pseudecheneis* and *Pseudolaguvia.* Among many other exotic flora and fauna, the region has a rich diversity of ornamental fish found in freshwater. There are about 267 different species of ornamental fish that are found in northeast India.

Among the northeastern states, Assam has the largest number of ornamental fish species (217), followed by Arunachal Pradesh (167), Meghalaya (165), Tripura (134), Manipur (121), Nagaland (68), Mizoram (48) and Sikkim (29). Ornamental fishes live in different water bodies in the different ecosystem from cold water to warm water in India's northeast region. Ponds, paddy fields, channels, wetlands, rivers are the main habitat of native ornamental fishes in this region. Lakhimpur district, on the foothills of eastern Himalaya of Arunachal Pradesh in Assam's northeastern corner, has been an ideal habitat for many ornamental fishes found in nature.

Fig. 11: Ornamental prawns

Conservation of the indigenous fish population

Taking into account the critical need to conserve fish genetic diversity of the region, rearing and captive breeding protocol for threatened and vulnerable species should be developed without further delay. Even though, there is great demand for the indigenous fishes like *Wallago attu, Ompok bimaculatus, Clarias batrachus, Heteropneustes fossilis* and *Channa* species, other commercially valuable fishes are in trace quantities. Other than these fishes, *Botia* spp., *Labeo* spp., *Mastacembelus armatus, Macrognathus* spp., *Trichogaster* spp. and *Mystus* spp. have tremendous potential as aquarium fish and their natural stock need to be strengthened either by external input or improving their breeding grounds.

3

Ornamental Fish Keeping Systems

Design and construction of farm including management is important with the advancement of cultural practices and increased involment of people. These are some common resources for breeding, rearing and display of ornamental fish.

Small scale commercial farms

Mazla/earthen vats

These are baked earthen containers and usually round in shape. These are used to keep the gravid female of livebearer fish for breeding. Its water holding capacity is 7.5-10.0 litre.

Fig. 12: Earthen containers for the breeding of livebearers

Earthen tanks/pits

These are dig-out tanks, where the fishes get a natural environment. Series of digout ponds may be constructed with proper shed nets and permanent sheds also. A boundary of nylon webbings should be provided around the tanks to prevent the entry of snakes, toads etc. Slope of the tanks should not be steep.

It is advisable to layout the plastic sheet (120-150 GSM) all along the pit. On the plastic sheets a layer of mud and cow-dung should be provided with a view to provide a condition of a natural environment.Size of one pit may be 20'x10'x4' with sufficient water supply arrangement and drain out system as well. Local varieties of fishes like Anabas, Barbs, Loaches, Danio, Eels, Gobies, Glass fishes may be reared up in these structures.

Fig. 13: Pits for small scale fish culture

Fig. 14: Cemented tanks for ornamental fish culture (In house)

Cemented Cisterns

Cemented cisterns are widely used in ornamental farms and it has been found that they give very good results. The size of the tanks may vary. Generally, a series of cemented tanks are constructed keeping space for farmers' movement along the tanks.

Normal sizes are – 8'x3'x2'; 8'x4'x2½'; 6'x3'x1½' to 2'; Even 5'x2½'x1½'.

Fig. 15: Cemented tanks for ornamental fish culture (Outdoor)

Fig. 16: Fiber tanks for ornamental fish culture

Fig. 17: Hapa breeding of ornamental fish in community ponds

Fig. 18: An indigenous ornamental fish rearing unit

Requirements for Breeding & Rearing Units

1. Deep Tube-well (700'– 1000')
2. Pipe-lines
3. Hatchery Shed
4. Covers with nets
5. Aerator/Air pump
6. Light arrangement
7. Breeding & Rearing equipment
8. Medicines

Some Pond Breeding Ornamental Fishes

Swordtails (*Xiphophorus helleri*)

a) Single tail swords
b) Double tail swords

- Platies (*Xiphophorus maculatus*)
- Guppies (*Poecilia reticulata*)
- Gouramis (*O. gourami/Helostoma temmincki*)
- Fighters (*Betta spleridens*)
- Angels (*Pterophyllum scalare*)
- Tetras (*Hyphessobrycon serape*)
- Barbs (*Puntius tetrazona*)
- Gold Fishes (*Carassius auratus*)
- Cichlids (*Cichlasoma carpintis*)
- Mollies (*Poecillia latipina*)

Re-circulatory system

The re-circulatory system is a closed system, which treats water by filtration in the fish tank. In this system, water is continuously exchanged from the fish tank to assure optimum growing environmental condition to prevail. Water is pumped into the tanks, through biological and mechanical filtration systems and then again returned into the tanks. There is no complete water exchange, instead only 5% to 10% water exchange rate per day is being done, depending on stocking and feeding rates.

Importance of recirculating system

It ensures the availability of good quality water for the culture of fish, that too at a low cost and provision of water conservation. It takes care of pollution as there are control over effluents and provision of effluent treatment. It permits the culture of aquatic organisms outside of the natural range. It can be practiced even with a limited source of water. It reduces the requirement of the land for culture and encourages multiple uses of waste materials like water and fertilizer. The threat of pollution is minimized with conservation of land and water.

Principle of Re-circulatory systems

A recirculation system involves fish tanks, filtration units and water treatment systems. The fish are housed in tanks and the water is exchanged continuously @ 5% to 10% per day depending on stocking and feeding rates through biological and mechanical filtration systems and then returned into the tanks.

System Components

- Source of Aeration – Fish requires optimum levels of oxygen to survive. Oxygen can be added to the system via an oxygen generator, to maintain suitable oxygen levels at high stocking rates. Sufficient oxygen and circulation of water will be maintained through aeration pumps.
- Tanks –The volume of most commonly used tanks is 5,000 to 10,000 litres. Non-corrosive plastic or fiber glass is recommended as tank material. Smooth round tanks with a conical shaped bottom draining are suitable.
- Pumps and Pipes – To move water in the system, there is a requirement of pumps and pipes.
- Mechanical Filtration
- The suspended solids should be removed.
- Pipes and equipment components should not become clogged with waste material.
- Some types include drum filtration, screen filtration, foam fractionation, settlement tanks, sand filters.
- Regular back-flushing requires preventing the accumulation of sludge.
- Biological Filtration-Fish produce ammonia and nitrites as metabolic waste products which are toxic. Bio-filters consist of a medium with a large surface area upon which nitrifying bacteria will colonize after a few weeks. These bacteria will convert toxic ammonia and nitrites into non-toxic nitrates by

oxidation. This process is known as nitrification. It usually takes a few weeks to a month before nitrifying bacteria colonize and the bio-filter becomes active.

- Sterilization-Ultraviolet sterilizer filters destroy the presence of diseases causing pathogens, parasites etc. Electric submersion heaters/coolers can be used for heat exchange.
- The artificial lighting is provided to give the fish an impression of a day/night regime. This is to stimulate normal feeding and behaviour patterns of the fish to ensure optimum growth.

Water Quality Management

Temperature

It is important to maintain water temperature because fish are less stressed when kept at their optimum temperatures and therefore become less prone to disease. To maintain the temperature, artificial heating or cooling devices may be used.

Oxygen

Dissolved oxygen is the most critical water quality variable and depends on water temperatures, stocking and feeding rates and the effectiveness of the aeration installed within the recirculation system. Dissolved oxygen concentrations about @ 5ppm ensure optimum survival and growth of the fish.

pH

The pH is the measure of the hydrogen ion (H+) concentration in the water. Optimum pH range is between 6.5 to 9 It may vary slightly depending on the culture species.

Carbon dioxide

Carbon dioxide is produced by the respiration of fish and bacteria within the system. Its high level may initiate respiratory problems as it will interfere with oxygen uptake and cause stress to the fish. The high value of carbon dioxide concentrations within the water column can also cause pH levels to decrease.

Stock Management

Culture stock must be managed carefully to ensure the health, survival and growth of the fish. Stock management consists of - Breeding technology, availability of fingerlings schooling fish, culture technology, growth rates and market acceptability.

Grading

Grading is an important exercise when farming fish using a recirculating system. Fish grows at varying rates therefore it is necessary to separate the fast-growing fish to prevent cannibalism. The frequency of grading depends on the species of fish however grading is generally performed more frequently when the fish are younger.

Feeding

The Feeding rates depend on the specific species. Manual labour is less in automatic feeders than manual feeding. Automatic feeders used for feeding zooplankton in the larval stage and dry pellets to fingerlings, whereas, in belt feeders, conveyor belt drops pellets into the water at regular intervals. Demand feeders triggering the feed disposal when required.

Disease and Stress

In re-circulatory systems fish are held at high densities therefore, they are more at risk to be stressed and prone to disease. It is important to monitor fish health continuously as if a disease outbreak does occur it can spread rapidly throughout the culture tank. Fish farmers usually ensure that tanks can be isolated from each other to prevent the disease from spreading the entire system.

Harvesting and Purging

Harvesting is mostly done by dragged nets in tanks. Care must be taken to prevent stress on the fish. Before stocking the fish, tanks must be filled with clean and freshwater. This ensures that the "off-flavour" that is often associated with farmed fish is removed; it is required to increase their market acceptance.

Advantages

- Growing season under control.
- Labor costs less due to automation
- Stock security from predators, disease, theft, pollution
- Less dependency on natural resources
- Controlled production
- Controlled marketing

Disadvantages

- Construction costs high
- Production costs high

- Complexity due to electrical, mechanical, hydraulic equipment
- Trained operators required
- Electrical reliability and back-up components

Aquaponics is the modified re-circulatory system to achieve higher productivity. It is a fish production system, in which water from an aquaculture system is fed to a hydroponic system where the by-products are broken down by nitrogen-fixing bacteria into nitrates and nitrites and are utilized by the plants as nutrients. The water is then circulated back to the aquaculture system.

Glass aquarium

Design & construction of aquarium tanks

Hobbyists and entrepreneurs can construct an aquarium as per their choice. The detail of materials required and the method to construct and maintain an aquarium is dealt here for hobbyists.

1. *Glass Aquarium*-(4'x1.5'x2')

a) Frame type

b) All glass types

Aquariums or Fiber-based tanks are arranged on aquarium racks, fitted with the arrangement of water supply pipelines, electricity, heater, air pumps, pH meter, hand nets, medicines, feeding bottles, weeds etc. Acrylic stands and aquarium covers are now available. These tanks & aquariums should have over-head shed/cover to prevent unusual rain and sun-heat etc.

General Sizes of Aquarium

1. 10" x 10" x 10"
2. 18" x 12" x 12"
3. 36" x 15" x 15"
4. 60" x 18" x 18"

Putty to prevent leakage

Homemade:

Fine Sand - 40 g

Lithaze - 40 g

Materials required for fabrication and decoration aquarium

Following materials will be required for fabrication and decoration of an aquarium:

1. Glass tank
2. Hood
3. Gravels and stone chips
4. Sand
5. Aquatic plants (natural and artificial especially meant for aquarium can also be used)
6. Color posters for background
7. Aquarium toys
8. Aerator
9. Air stone
10. Thermometer
11. Thermostat
12. Filtration unit
13. Water (clean and chlorine-free water)
14. Ornamental fish as per choice
15. Artificial food
16. Hand net
17. Bucket and mug
18. Sponge

Construction and fabrication of glass aquarium

Fish bowl is the simplest tank to keep the ornamental fishes. It is made up of all-glass having various capacities generally below 5 liters. It serves to accommodate one or a few fishes only.

An aquarium tank is commonly used for keeping ornamental fishes.Its sides and bottom are made up of glass, fitted into a metal frame of either angle iron or aluminium by means of bitumen. Depending upon the size, an aquarium tank can be classified as small, medium or large. A small aquarium of 45 X 25 X 25 cm is good for small species and looks very attractive.The medium aquarium of 60 X 30 X 30 cm is the most popular one. The large aquarium 90 X 30 X 30 cm or 90 X 45 X 45 cm is also popular and is a standard size for home.

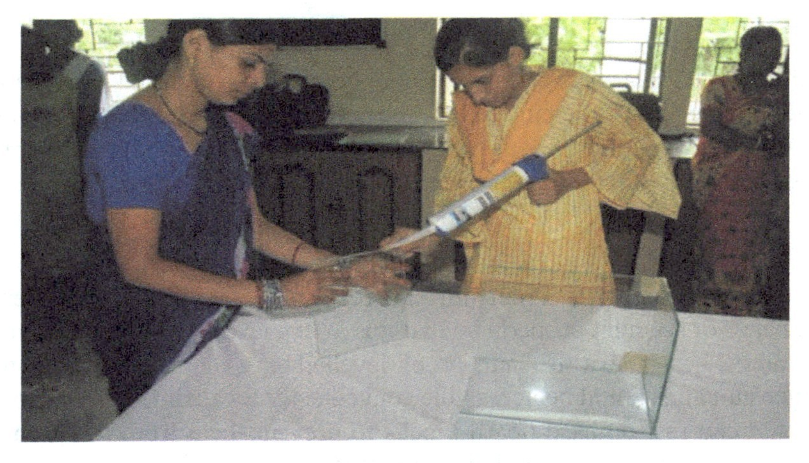

Fig. 19: The making of aquarium

Fig. 20: A modern aquarium

Other parts of an aquarium

Glass cover

A glass cover is must to prevent the escape of the active fish from jumping out of the tank. It also reduces the evaporation and protects the light fixture from water splashing. The cover, however, is designed to permit the free exchange of the air above the water with the surrounding atmosphere.

Hood

In addition to the glass cover, an aquarium hood is also required. It is made as per the specific size of the tank to fit lamps and make provision for feeding the fish. A reflector for lamps can be fitted in the hood and the starter for the fluorescent lamps should only be mounted on the hood.

Aquarium stands

The stand on which the aquarium can rest should be decided as per the convenience of seeing it properly so that it is at eye level. A full filled aquarium with water is heavy, so a firm and strong-levelled base should be selected to set the aquarium and should be accessible conveniently for maintenance purposes. The stand can rest on hard board blocks to prevent them from damaging the floor covering if a stand is used on a carpeted floor. After keeping the stand at a suitable place, 1-3 cm thick thermocol or polythene sheet should be spread to give a cushion to the supporting surface of the aquarium.

Placing of an aquarium

The aquarium should be kept on a flat place and firm surface or a table. Stands made up of iron or wood can be used for keeping an aquarium.The aquarium should be fixed at a height parallel to the viewer's eyes so that the fishes can be seen easily and one can enjoy their activities. Precaution should be taken that direct sunlight is not falling on the aquarium because it will boost algal growth and glasses of the aquarium will not look clean. The aquarium should be placed near the electricity connection to provide electrification and aeration of the aquarium easily.

Installation of an aquarium

The aquarium should be installed after checking for any water leakage. Then after washing it thoroughly with salt water, it should be wipe carefully with a clean cloth. Use of detergents or disinfectants, should be avoided as these are likely to prove toxic if any residues still remain after rinsing the tank. Gravel or sand can be used to cover the floor washing of the covering material thoroughly with clean water is essential. A rinse with potassium permanganate is always good as prophylactic measures to restrict the growth of pathogens. If an underwater filter is used it should be installed before adding the gravel. A natural-looking scene can be created as per one's own choice by using gravel and sand, without blocking the view of the fish. The sandy bed is good for easy rooting of ornamental plants. While spreading the gravel. care should be taken to spread it evenly, but slightly deeper at the back than front for an attractive look.

Setting of aquatic ornamental plants in the aquarium

Aquaniusm are decorated by planting ornamental plants for beauty and the well being of the fishes. In addition to decoration, they provide shelter during day time and remove the harmful gas (Carbon dioxide) given off by the fish in the process of breathing. They are capable of producing oxygen also.Although artificial plastic plants are now available and look extremely realistic, the provision of living plants in the aquarium will undoubtedly help to maintain the biological balance of the environment.

A wide variety of plants are available in the market to decorate the aquariums. They are *Cryptocome, Echinodorus, Elodea, Hydrophyla, Ludwigia, Cabomba, Myriophyllum, Vallisneria*, etc. These plants may be soft leafed or stout leafed. The soft-leafed species are good food for herbivorous fishes while the firm stout-leafed plants are favourable either for spawning or as nest-building materials.These plants can be planted directly in the sand bed or in the containers specially made for this purpose. The plants should be disinfected to kill harmful germs, parasites and other enemies of fish. Disinfection can be done by giving a dip to them in 0.1% potassium permanganate solution for 10-20 minutes and then should be washed in running water properly. At the rear side of the aquarium the taller plants should be planted while broader leafed plants at the centre and smaller plants at the front is advicable. Plants are best rooted in the aquarium when it is half-filled.

Decoration of an aquarium

Rocks provide a hiding place for some of the fishes besides giving beauty to the aquarium. The rocks selected must be inert like granite, flint, slate, sandstone and well-washed coal. Sea shells and corals will dissolve in a freshwater aquarium, so it should not be used. Natural rocks may be collected or can be bought from the pet shops as per the choice to decorate the aquarium. Generally, tall pieces are placed at the back and small pieces at the front. Proper cleaning is necessary before placing it inside the tank. Wood in different forms can also be decorative in the aquarium and it should be used only after performing all the required cleaning procedures.

Apart from rocks and wood a number of decorative toys like plastic bubblers in the guise of a mermaid, underwater diver, oyster shells, angler, human skull, tortoise, frog etc. can be placed for decorative purpose.

Fig. 21: Different types of toys for aquarium decoration

Light Arrangement

Proper illumination is not only important for an attractive look but also important for the healthy functioning of the home aquarium. It is very necessary for plant growth and fish activity in the aquarium besides seeing the beauty of fishes and enjoying the aquarium in habitats. However, too much light enhances the growth of algae and too less light retards the growth of plants. The lighting with a 40-Watt fluorescent bulb or tube light for 10 to 12 hours per day is ideal for the normal house aquarium. Nowadays, specially designed fluorescent tube lights are available in the market for attractive illumination and displaying fishes to their best effect.

Aeration and heating arrangement

The prime aim of aeration is to oxygenate the water; it also helps to circulate the water, which helps to maintain a constant temperature throughout the aquarium. The diaphragm pump is a popular type of areator used for aquarium aeration. It is operated by a vibrating electromagnet, which activates a diaphragm for drawing air. Air is conveyed from the pump through narrow-bore transparent plastic tubing and is distributed by control valves. Due to continuous operation, their coil gets heated up followed by the diaphragm reducing the efficiency of the pump. The aerator should be kept at a higher level than the water of the aquarium; otherwise, water will creep into it when switched off. The air stones (micropore airstone) are connected to air tubes from the pump for providing minute air bubbles that diffuse more oxygen in the water. Sudden temperature fluctuation results in mortality of fishes.

The tropical aquarium fishes need temperature in the range of 22-30^{0}C and to maintain the required temperature during winter months it is necessary to arrange heating equipment that can be fitted in each aquarium. A water heater of 5/10 watts capacity is required for one gallon of water to maintain the tropical fishes during the winter months. Basically, the heating equipment (submerged in the water) of the aquarium is an electric heating coil added with a thermostat to control it. Both are housed in a glass tube. A thermometer is essential to keep inside the aquarium to monitor the temperature. It is advisable to place it at the front where it can be easily monitored. It is best that arrange the heater and thermostat at opposite corners of the tank to get correct temperature readings of the aquarium water.

Fig. 22: Accessories and their placement in aquarium

Filling an aquarium with water

Pouring the water directly to the aquarium may cause a stirring of the sand and gravel resulting turbid water. To overcome the problem, a shallow pot like an inverted bowl should be kept on the sand and then gently water should be pour on that. It flows gently over the brim and fill the tank without disturbing the sand and gravels.After setting up the aquarium, it should be left for 24 hours with aeration before introducing the fish for the satisfactory result. It also help in checking the working of the equipment like filter, aerator and toys etc.

Acclimatization before stocking of fish in the aquarium

After arranging everything in the aquarium properly, it is time to stock the fish. Selection of the fish species is very important for keeping them in the aquarium. Generally, the beginners select locally available hardy fishes, which require less care. The home aquarium is a single species tank where only one type of fish species is kept. Some fishes are of the habit of attacking others and are kept in this type of tanks. Majority of the fish tanks are of the community type where diffrent varieties and sizes of compatible types of fishes are kept together to give a good decorative look. It is advicable to buy and release one or two pairs of fish in each week until the aquarium reached its maximum capacity.

At the time of addition of a new fish, the aquarium light should be turned off and the fish should be conditioned by floating the plastic bag for 20-30 minutes on the surface. After opening the bag fish should be allowed in the aquarium water to flow in and move very gently. Then lights should be turned on and a small quantity of food should be given to distract their attention from the newcomers.

Number of fishes to be kept in an aquarium

There are various types of ornamental fishes available for hobbyists. The selection and combination of fishes in the aquarium depends upon their availability and compatibility in the group. It's advisable to select small fishes for aquarium. The common species are black molly, platy, guppy, swordtail, fighter, angel barb, goldfish etc. In addition to these fishes, any fish having ornamental in nature and suitable for captive rearing can be selected for aquarium. Their behaviour with other fishes in the group should be studied before keeping them together in an aquarium for rearing. Fishes of the small size of 2-3 cm. are advised for rearing in an aquarium. A space of about 50 sq. cm for a fish of size 2.5 cm is recommended for the best rearing. Considering this recommendation 35 fishes of size 2.5 cm are allowed to keep in a 60 × 30 cm size aquarium having 1800 sq.cm water surface.

Feeding and Maintenance

Fishes are of varying feeding habits in nature. Some are carnivores; attack others for their food. But all require live food for their proper growth. They also can take artificial diets in the absence of natural food. Different types of ornamental fish feeds are available in pet shops. As a general rule, the fishes can be fed twice daily in the morning and evening. The quantity of feed depends on the consumption of fish in half an hour. Fish should be fed with a variety of feeds, like feed containing vegetable matter, live feed, dried feed etc. a two, three or even five times per day, without exceeding the limitation on quantity to be fed.

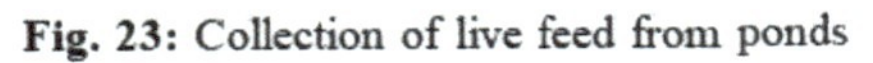

Fig. 23: Collection of live feed from ponds

Fig. 24: Artificial feed

Materials and specifications

1. **Hand-net:** A hand- net is generally required for catching a fish for inspection and for transfer to elsewhere.
2. **Algae cleaner:** To clear the glass of the aquarium of algae, a fine steel wool held in the hand can serve the purpose.
3. **Siphon pipe:** This is used to siphon water out of the aquarium at the time of exchanging water.
4. **Water testing kit:** Water-testing kit is used to test the nature of the water from time to time, such as pH, dissolved oxygen content, free carbon dioxide, etc.
5. **Aquarium toys:** The toys and some of the equipment, which has the algae and other sediment attachments, can also be cleaned.
6. **Aquarium plants:** Plants have to be trimmed and decaying leaves have to be removed.

   The working efficiency of all the equipment has to be rechecked for its efficiency.

Maintenance of Aquarium

Water from a well-stabilized aquarium need not to be changed at all. But due to many factors like over feeding, high stocking density, less oxygen supply, improper lighting, accumulation of waste materials, disease etc. the water in the aquarium becomes unhygienic to the fishes and plants. Fishes start surfacing and die in a later stage. The condition makes necessary for a partial or complete change of water. A partial change of water at least 10-30% of the water on weekly or fortnightly intervals is enough for a home aquarium. A polythene tube with an adapter having perforated ends can be used for siphoning the bottom sediments during this activity.

Once an aquarium is set with all equipment and stocked with fish, due to regular feeding and metabolic activity of the fish the water condition is liable to change over a period of time and it requires regular maintenance. Care should be taken for the following points to maintain a home aquariums.

1. The visibility through the glass panel may get reduced due to accumulation of feed waste, slimes, bacteria etc. Therefore, regular wiping even on a daily basis is the necessity. A damp cloth can be used to clean it without scratching the glass. Now-a-day's magnetic glass-cleaning devices are available in the market for this purpose.

2. The health and condition of the fishes in the aquarium have to be observed daily. It will be always better to count the numbers every day during morning feeding.
3. The algae and other materials build up on the cover glass should be cleaned during the regular maintenance.
4. A tank once filled with water should not be moved as it may start a leak or even break.
5. During the maintenance of different aquarium equipment, the working principles and manufacturer's instructions have to be considered for getting the optimum efficiency of the equipment.

Table 2. Aquarium dimensions volume & weight.

Length (cm)	Width (cm)	Height (cm)	Gross volume (litre)	Net volume (litre)	Weight (kg)
80	30	40	96	77	125
80	40	45	144	115	194
100	40	50	200	160	280
120	45	50	270	216	400
150	50	50	375	300	550
200	50	50	500	400	700
250	60	60	900	800	1000

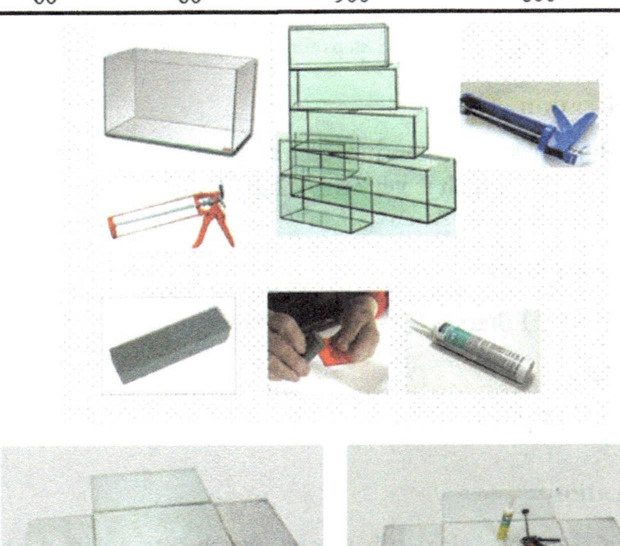

Fig. 25: Requirements and steps for making a glass aquarium

Mechanical Filtration

Most filter media serve to filter mechanically the water to some degree. Mechanical filter media, which is very fine, trap greater quantities of debris and plug more rapidly.

- Clip-On Power Filters
- AquaClear Power Filters
- Canister Filters
- Elite Hush Power Filters

Chemical Filtration

With the help of some chemical substances, water gets purified. It improves the water quality by eliminating the odour, removing chlorine, eliminating medications after disease treatments, neutralizing heavy metal ions and effecting changes in hardness and pH levels.

Biological Filtration

It is accomplished by various beneficial strains of bacteria like *Nitrosomonas* and *Nitrobacter.* Nitrifying bacteria utilize toxic nitrogenous compounds, ammonia and nitrite as their energy source, and produce nitrate, relatively less harm by product.

Heaters

To maintain the temperature of water in aquariums, heaters are required. The types include glass immersion heaters and under gravel heating. There are also heating mats that may be placed under the aquarium. Heater submersible with easy read temperature settings, easy grasp and adjust, temperature set dial of electronic calibrated control for accurate temperature settings are preferred.

Testing of Water

It is important to monitor the quality of aquarium water on a regular basis.Tap water contains heavy metals from the piping it runs through, and chlorine or chloramine, so need to condition tap water before filling, changing or replacing water in the tank.

Fig. 26: Test kits to measure water parameters

4

Ornamental Plants

Ornamental Plants are a real attraction to the aquarium tank.While some plants are both ornamental and functional, people usually use the term "ornamental plants" to refer to plants which have no value beyond being attractive, although many people feel that the attractive value itself is quite enough. They provide optimal habitat for the fishes especially for the fry (baby fishes). It maintains the ecosystem of the aquarium tanks by providing good oxygen, food and also acts as a good hiding place for the baby fish. Mostly all aquatic plants are very easy to grow and care. They are always good for gold fish and other similar fishes. There is a huge variety of plants. Ornamental plants are grown for decoration, rather than food or raw materials. They are most often intentionally planted for aesthetic appeal. However, ornamental plants also serve some less obvious uses such as for the purpose of fragrance, for attracting wildlife and for cleaning the air. Ornamentals encompass a wide array of plants. Commonly, ornamental plants are grown for the display of aesthetic features including: flowers, leaves, scent, overall foliage texture, fruit, stem and bark, and aesthetic form. The concept of ornamental plants is used for decorative purposes in aquariums, gardens, home gardens, landscape design projects, squares, parks etc.

Aquatic plants predominantly grow in water. They vary greatly in type, with some being quite similar to common land plants while others are quite different. Aquatic plants can be commonly classified into four types: algae, floating plants, submerged plants and emerged plants. Aquatic plants live in water. This means that they require being submersed to survive or that they can only grow and thrive in water. Many aquatic plants can tolerate being emerged, meaning they're out of the water for short periods of time, but require immersion in water for long-term survival.

There are some plants that simply can't tolerate being emerged at all, but this is rare. These plants can only live when fully immersed in water. Most macro-algae require total submersion in water for the span of its entire life. Most aquatic plants, however, have roots underwater but can only grow or flower in a partially emerged state. All plants, aquatic and otherwise, require sunlight, soil, gases and water to survive. Plants make their own food via the process of photosynthesis; this means that they need a source of energy to drive this biochemical process, and the sun provides a perfect one. Plants can carry on for periods without sunlight, just as animals can live on stored fuel for a while in times of need. The soil offers a place for the plant's roots to take hold.

Air contains large amounts of the carbon dioxide gas (CO_2), plants need to power photosynthesis, but aquatic plants have evolved to draw in CO_2 that is dissolved in water in relatively small amounts. Finally, plants need water to combine with the CO_2 to complete photosynthesis by generating oxygen and glucose.

Types of Aquatic Plants

There are 6 types of aquatic plants, based upon morphology:

- **Amphiphytes:** This type of plants can live either submerged or on land.
- **Elodeids:** This type lives submerged apart of their flowers which come above the water line.
- **Isoetids:** Isoetids plants entirely remain submerged for their full life cycle. It can be found on wet lands having low nutrient availability. It has a slow growth rate.
- **Helophyte:** Helophyte plants grow partly submerged. Often rooted in bottom but leaves grow above water line.
- **Nymphaeids:** These aqua plants rooted to bottom but leaves float on water surface.
- **Pleuston:** Water plants which freely float on water surface.

Algae

Algae are morphologically simple, chlorophyll-containing organisms that range from microscopic and unicellular (single-celled) to very large and multicellular. Algae are the oldest and the most common type of aquatic plant. They are found primarily in the ocean and they have no roots, stems or leaves. Algae are extremely small but they are the basis of the aquatic food chain. They belong to three different groups, recognized since the mid-nineteenth century on the basis of thallus color: red algae (phylum Rhodophyta), brown algae (phylum Ochrophyta: class Phaeophyceae), and green algae (phylum Chlorophyta). Examples of algae include *Lyngbya, Cladophora, Spirogyra* and *Oedogonium* etc.

Floating-Leaved Plants

Floating-leaved macrophytes have root systems attached to the substrate or bottom of the body of water and with leaves that float on the water surface. Common floating leaved macrophytes are water lilies (family Nymphaeaceae), pond weeds (family Potamogetonaceae). Floating plants can be found in fresh or salt water. The leaves of these plants are firm and remain flat in order to absorb more sunlight. Common examples of floating plants include various types of lilies (such as the water lily or banana lily) and the water hyacinth.

Submerged Plants

Submerged plants are rooted plants with flaccid or limp stems and most of their vegetative mass is below the water surface, although small portions may stick above the water. The leaves of these plants are thin and narrow. Submerged water plants can be divided into two main types:

- Underwater plants that grow with their foliage totally submerged underwater, such as Eel grass (*Vallisneria*).
- Emergent water plants with foliage both under the water and also on or above the water surface, such as water milfoil (*Myriophyllum* species).

Emerged Plants

Emerged plants are rooted plants often along the shoreline that stand above the surface of the water (cat tails). The stems of emergent plants are somewhat stiff or firm. Most of their vegetation exists above water. These plants need constant exposure to sunlight. Examples of emerged plants include knot weed and red root.

Fig. 27: Ornamental plant in aquarium

Fig. 28: Ornamental plants in local market

Plants are used to give the aquarium a natural appearance, oxygenate the water, act as feed and provide habitat for fish, especially fry (babies) and invertebrates.

The freshwater aquarium plants provide natural filtration for the water, help keep fish healthy, and can even help to breed the fish.

Uses of Freshwater Aquarium plants

These plants act as:

- Natural aquarium carpet
- Hiding places for fish
- Concealing aquarium fixtures
- Balance the aquarium environment

Suitability of Ornamental Plants

The plants chosen for the aquarium must be hardy and need little maintenance. Green plants don't require as much direct sunlight. Excessive sunlight may cause algal growth which will negatively impact the aquarium environment. It could result in the need for more maintenance and cleaning.

There are many ornamental terrestrials or tropical marginal plants sold as aquatic plants. These plants may look good and can do fine completely submerged for a while, but these plants will eventually drown and should not be used. It is very important to select true aquatic species to prevent unnecessary disappointment and loss.

Best Plants for Freshwater Aquarium

Water Wisteria

Water Wisteria grows fast, quickly adding depth and beauty to the freshwater aquarium. It needs minimal maintenance, which makes it a fuss-free plant to have. This plant has lace like green leaves which requires minimal light. It forms a carpet on the floor of the aquarium. There are many species of the plants, having wide distribution.

Fig. 29: Wisteria

Java moss

It is quite popular for freshwater aquariums. It requires low-maintenance and tends to grow as fuzzy green plants on the bottom of the aquarium. It provides protection to baby fish and acts as actual secondary food source for

Fig. 30: Java Moss

baby fish. It does not require any special attention. It accepts all kinds of water, even weak brackishwater, and all kinds of light qualities. It grows best at 21 to 24°C, but can live in temperatures of up to 29 to 32°C. It makes a good foreground plant.

Nelumbo

Fig. 31: Nelumbo

Nelumbo nucifera, also known as Indian lotus, sacred lotus, bean of India, Egyptian bean or simply lotus, is one of two extant species of aquatic plant in the family Nelumbonaceae. It is often colloquially called a water lily. Under favorable circumstances the seeds of this aquatic perennial may remain viable for many years, with the oldest recorded lotus germination being from seeds 1,300 years old, recovered from a dry lake bed in north eastern China. It has a very wide native distribution, ranging from central and northern India (at altitudes up to 1,400 m or 4,600 ft in the southern Himalayas), through northern Indo-China and East Asia.

Victoria

Fig. 32: Victoria

Victoria is a genus of water-lilies, in the plant family Nymphaeaceae, with very large green leaves that lie flat on the water's surface. *Victoria amazonica* has a leaf that is up to 3 metres (9.8 ft) in diameter, on a stalk up to 8 metres (26 ft) in length. The genus name was given in honour of Queen Victoria of the United Kingdom.

Lilaeopsis

Fig. 33: Lilaeopsis

The *Lilaeopsis brasiliensis*, commonly known as Brazilian micro sword, is a short-stemmed carpet-like plant commonly used in the foreground of many tanks. It can be found growing partially or fully submersed along side riverbanks and streams all around Brazil, Argentina, and Paraguay. It is a petite growing plant, reaching a maximum of 3 inches in height, which makes it suitable for many tanks. Nevertheless, its growth can be difficult or at least tricky, so beginner aquarists shouldn't attempt to grow this plant until basic knowledge about growing aquatic plants is obtained. This plant can be kept in all tanks due to its small size. It will grow well in soft to moderately hard water with a slightly alkaline pH between 6.8 and 7.5. Water temperature should be between 70 to 83 degrees Fahrenheit, so it can be planted in a wide range of habitats. Light is very important for the Lilaeopsis to thrive in the tank. Lighting with a minimum of 3 watts per gallon is necessary for its survival and it should be direct light.

Hydrilla

Hydrilla, otherwise known as water weed, water thyme or freshwater seaweed is a very popular aquatic plant that is distributed world wide. It has pointed, bright green leaves about 5/8 inches long. it grows horizontally along the bottom of the water body. Side shoots and new tubers can develop at the nodes as the plant grows. As the water temperature increases, the stems elongate, sending the shoot tips toward the water surface, creating a thick mat of vegetation. A very good food source for herbivore fish. In aquarium it would be wise to check them frequently and cut them off if they grow too long otherwise they will overpopulate the aquarium. Usually, it would be unwise to put Hydrilla in tanks that contains goldfish, carp or young giant gourmai because they will 'trim' all the leaves; they will look ugly. But they won't die under such condition.

Fig. 34: Hydrilla

Soft Hornwort

Soft hornwort is a herbaceous species of plants in the family Ceratophyllaceae, with a self-supporting growth habit. It is associated with freshwater habitat. It has compound, broad leaves. It is a photoautotroph. It is another popular aquarium plant. It's a very soft and delicate plant which is loved by plant eating fish. Scientific name is *Ceratophyllum submersum*. This plant was first reported from Europe, Central Asia, and northern Africa but now can be found everywhere. It is a submerged, free floating bright green aquatic plant having a feathery appearance in water. The leaves are soft.

Fig. 35: Soft Hornwort

Mayaca

It can be found in swamps and wet lands of Southern and Central Amercia. This is another great Aquatic plant suitable for Tetra or Barb tank. It needs sun light to grow healthy. *Mayaca fluviatilis* is a stem plant from South and Central America. The stem is extremely soft and flexible which, in addition to the very narrow, bright green leaves, gives the plant a light and graceful expression. It branches willingly and the rapid growth makes regular pruning necessary.

Fig. 36: *Mayacafluviatilis*

This plant prefers soft, slightly acidic water, but it also thrives well in medium hard water. Mayaca requires moderate light or the lower parts of the stems can start to turn yellow and die off. It is propagated through cuttings that easily take root in the substrate. Mayaca can be grown submersed or emersed. Propagation is simple and straight forward.

Amazon Sword

With their blade like appearanc and impressive endurance, the Amazon Sword plant has earned a spot among the most popular aquatic plants. An Amazon Sword makes a great background plant. When planted alone, it can be an eye-catching center piece plant. Its lush green leaves will really stand out especially against a black background. When planted with others of its kind, an Amazon Sword Plant can create a thick green "forest-like" effect. Either way, an Amazon Sword Plant can be just the right plant to hide a power filter intake tube or an aquarium heater. This plant is beautiful and easy to care for, and will create a forest like effect in your tank – it's great for beginners and experienced aquarists alike. Goldfish can be rough on sword plants too, so they may not be the best plants for a Goldfish tank.

Fig. 37: Amazon Sword

Oriental Sword

It is very much similar to Amazon Sword plant but leaves are little bit different. Its Scientific name *is Echinodorus oriental*. It is commonly known as the Oriental Sword, and is one of the most interesting sword varieties to enter the

Fig. 38: Oriental sword

trade in the past 20 years. *E. oriental* propagates most often by side shoots coming off the rhizome. They have also been observed with adventitious plants on the peduncle. A very hardy plant should be even better than Amazon Sword for a Goldfish tank. This plant also could grow very tall, even leaves coming above water.

Anubias nana

Fig. 39: *Anubias nana*

Anubias nana is one of the more appealing midground plants. It tolerates nearly any water quality or environment. *A. nana* is a short plant with broad leaves. Its dark green colors make it an attractive plant and it will help keep your tank water clean and oxygenated. With curved stems and large semi-round leaves, it's a great match for the stone aquascaping present in most aquariums. It has curved stems with medium-sized, semi-round leaves. It grows best in water of temperature between 22 and 26 degree Celsius. Growth is optimal in medium lighting. It is decorative, protective and looks beautiful in any aquarium placement.

Pygmy chain sword

Fig. 40: Pygmy chain sword

This plant isn't often seen in aquascaping. Its unique look and adaptability have made the Pygmy Chain Sword plant immensely popular. Being a short plant, it makes a wonderful addition for the front of the aquarium or as a 'ground cover'. If proper care is taken, it rapidly reproduces to form a dense 1-2 inch (2.5-5.0 cm) tall "lawn". It's useful for placement around hardscapes, and is beautiful when properly trimmed. It looks strikingly similar to the most lawn grass. It grows best in the water temperature between 22 and 26ºC. Growth is optimal in medium-bright light. In addition to decoration and protection, it is good for placement around hardscapes.

Downoi

Fig. 41: *Pogostemon helferi*

Pogostemon helferi is an unusual, yet beautiful, foreground plant for aquariums. Its unique look is due to the curled appearance and dark green shade of its leaves. Thailand locals named this plant "daonoi," meaning, "little star." Besides having an interesting name, this is one of the most unique foreground plants available to aquascapers today. It has a striking zig-zag shape in its leaves, and grows in a 'blooming' pattern that's visually appealing in front of hardscapes. These are beautiful 'blooming' growth pattern. Zig-zag shaped leaves. It grows best in water having 22-26 degrees Celsius temperature. Growth is optimal in medium lighting.

Dwarf sagittaria

Dwarf sagittaria is an easily-maintained plant that maxes out at around 4-6 inches, making it perfect for midground aquascapes. Placing Dwarf Sagittaria around stonework or driftwood is an ideal location, giving it a perfect place to root into the wood or stone, and is an ideal complement. Vibrant green leaves with curved blades. It can grow both partially and fully submersed, so it can adapt to a wide range of conditions in the home aquarium. Growing dwarf sagittaria is relatively easy, and this is one of the few plants that can tolerate very high pH and hard water conditions. It should be planted in a nutrient rich substrate, or the water column should be regularly fertilized. It is especially sensitive to low levels of iron, and any yellowing of the leaves normally means an iron deficiency.

Fig. 42: Dwarf sagittaria

Maintenance of aquarium plants

For enjoying the right balance, plants care required like check on the health and wellness of fish, give a little attention to the aquatic plants as well. Here is a list of things to do:

Supply the right substrate

- Aquarium plants need substrate material, used to cover the bottom of an aquarium in which to anchor their roots. Plants can grow in most types of substrate, but two to three inches of laterite covered with an inch of gravel is ideal.
- It's possible to keep plants in the pots in which they were sold. However, "planting" plants in substrate provides for a more natural look, and is more conducive to root development.

Provide the right light

- Without proper lighting, plants won't survive. Plants need light for photosynthesis, a process in which they generate energy for growth. An added benefit of photosynthesis is that it produces oxygen for the aquatic life in aquarium.

- Full-spectrum, fluorescent lighting is a must. Make sure to give plants 10-12 hours of light per day. If using fluorescent lighting, remember to change light bulbs at least every 12 months, as their intensity may fade. If light fails to emit a full spectrum, plants won't thrive.

Use the right algae-reducing techniques

- Like a garden weed, an alga competes with aquarium plants for light and nutrients. There are several solutions for eliminating algae. Algaecide, a formula based on antibiotics or chemicals, kills the algae but may have adverse effects on aquarium.
- There are a variety of herbivorous aquatic lives that can help keep algae in check and also physically remove the algae using an algae scraper. For best results, alge should be scurbbed weekly and over feeding should be ovided, which may result algal growth again.

Use the right fertilizer

- Plant growth may be maintained by adding a fish-safe, iron-based fertilizer. Special slow-release fertilizers that are designed to aid in freshwater aquarium plant growth will be better Phosphate fertilizer, as algae thrive on phosphates should not use.

Practice the right "Aquascaping" skills

- Certain types of aquarium plants require pruning, especially tall stem plants like *Rotala indica*. These plants grow across the water's surface if not pruned back. When this happens, they can block precious light from other plants. It is important to remove dead leaves and dying plants, too. Decomposing matter will affect aquarium's water quality.

Fertilizers required for plants

Like fish, plants need care and maintenance. By adding the required nutrition like, CO_2, nitrogen, phosphorous, iron and micronutrients in different ways will result in optimal growth and beautiful plant colours inside the aquarium. The composition of nutrients is conveniently divided into macro and micro nutrients. Macro nutrients are needed in larger quantities while micro nutrients are sufficient in smaller quantities.

Macronutrients are calcium, sulfates, phosphates, potassium, chloride, sodium, nitrogen and magnesium. These nutrients are provided by fish and fish food in ample supply. Macro nutrients do not need to be added frequently, if at all, as they will be mostly replenished through water changes. Essential micro nutrients such as iron, manganese, zinc, boron, copper, cobalt, and molybdenum on the other hand have to be added frequently. The main function of these nutrients is the promotion of growth hormones, photosynthesis, cell development, plant metabolism, and nitrogen assimilation.

The assumption that plants take on most nutrients through their leaves is incorrect. Leafs absorb CO_2 and release oxygen. Essential nutrients such as iron, phosphates and nitrogen are readily absorbed by the roots under anoxic conditions, found in the substrate. Plant fertilizers are available as liquid or substrate fertilizers. Both should only contain the micro nutrients. Liquid fertilizers have to be dosed more frequently; substrate fertilizers last longer. Since there are no obvious differences in efficiency, it is up to the aquarists' preference which one to use.

Next to the micro nutrients, fertilizers contain chelating agents. Chelation is an organic molecule which binds metal ions thus protecting them from early precipitation. The preferred type is abbreviated DTPA because of its stability up to a pH level of 7.5. Unfortunately, some fertilizers contain the chelating agent EDTA, which is much cheaper. However, chelate EDTA is only stable at a pH up to 6.0 and therefore mostly having no use in aquariums.

Another important yet often overlooked aspect in using fertilizers as water conditioners. Many conditioners eliminate heavy metals and since many micro nutrients are metals, plants can be deprived of essential nutrients despite the frequent addition.

Table 3. Fertilizer for Aquarium Plants Available in the Market.

Sl. No.	Plant Fertilizers	Top Features	Pictures
1	Seachem Flourish	Contains a rich assortment of important micro elements, trace elements and other nutrients	
2	Api Leaf Zone Aquarium Plant Food	Contains chelated iron and potassium essential for lush green leaves	
3	Api Co_2 Booster	Adds essential carbon in a form plants can absorb from the water	
4	Extremely Easy to Use Liquid Fertilizer Thrive All in One Liquid Fertilizer	Provides the necessary macro and micro nutrients	
5	Liquid Npk+M - Aquarium Liquid Fertilizer	This product is tailored to the EI dosing method but can be adapted to other dosing techniques	
6	Plump Your Marimo With Luffy Fertilizer - Marimo Food Boosts Growth	Liquid plant food, especially for marimos. Contains exact ratio of Phosphate, Calcium and Nitrates in addition to required salts.	

Transportation of Aquatic Plants

There are many ways to transport aquarium plants. In the past, people used to roll them in old wet newspapers, then placed them in zip lock seal type plastic bags and put the bags in the plastic bucket. It should be sure to use appropriate bags and good rubber band to secure the top; those made specifically for shipping and transporting fish are also the best ones to use to transport or ship plants.

It's best to leave some air in the bags, but for short and extreme situations the bags can be compressed and ship them pretty much flat. It must remember, the plants were only in those conditions for less than 12 hours; for regular shipping be air should leave air in the bags. On the other hand, it should be sure that the bag isn't too filled with air. It should not be like a balloon or it may pop under different air pressure changes that may occur in shipping.

The insulated boxes do add peace of mind, and can be used them whenever possible. It must be ensured that appropriate padding is used. Packing peanuts and bubble wrap are good for packing around plants. It must be ensured that the plants aren't too tightly or loosely packed, as not required to squish them or have them flying around the box. One should remember that, unlike fish, in most cases plants can be shipped without water, as long as the leaves are damp and the plants are in sturdy plastic bags. This makes them much easier to ship.

Aquarium Plants for Entrepreneur

Incorporating ornamental ponds and containers of aquatic plants into backyard landscaping has become quite popular over the past ten years and many garden centers and home improvement stores are featuring a variety of equipment for homeowners. Pre-formed ponds as well as dug and lined ponds bring together various segments of landscaping with the use of stones, filters, pumps, lights, fish, and aquatic plants to provide attractive features. There are many varieties of ornamental aquatic plants and the market for these is growing steadily. Aquatic plants in some instances can be considered weeds and can be fouling organisms that need to be removed. Plants such as water hyacinth produce beautiful flowers, but these same plants may cause problems in some natural bodies of water. However, in a controlled cultured setting they grow well and propagate themselves with little assistance and serve as a decorative addition to backyard water gardens. Now days, aquarium plants is having good demand both in domestic as well as Industrial market. People if starts business of culturing and marketing of these plants can earn a lot, its, depends on the selected plants and location of the market. Therefore they should focus on e-marketing to covering the all parts of the globe. Major important points to be considered for becoming an entrepreneur in the field of aquarium plants business:

- Selection of right plants to culture and market – this can be done by listing the plants having good market demand and fit in the aquarium of particular ecological area
- Survey the area for the existing competition of plants business
- Making a plan for funding options and benefit & cost analysis
- Space required for Culture – Home terrace to farm yards can be used
- Marketing - Now-a-day there are many options, but e-marketing is the best option. One can work for the e-marketing options like Nursery live, Plants guru, My bageecha, Green wave nursery and Amazon etc.

Tissue Culture of Ornamental Plants

Tissue culture plants are also known as *in vitro* plants. Tissue culture plants are grown under laboratoy conditions. It's a collection of techniques used to maintain or grow plant cells, tissues or organs under sterile conditions on a nutrient culture medium of known composition. Plant tissue culture technology is being widely used for large scale plant multiplication. Apart from their use as a tool of research, plant tissue culture techniques have in recent years, become of major industrial importance in the area of plant' propagation, disease elimination, plant improvement and production of secondary metabolites. Small pieces of tissue (named explants) can be used to produce hundreds and thousands of plants in a continuous process. A single explant can be multiplied into several thousand plants in relatively short time period and space under controlled conditions, irrespective of the season and weather on a year round basis. Endangered, threatened and rare species have successfully been grown and conserved by micropropagation because of high coefficient of multiplication and small demands on number of initial plants and space. Aquatic plants from tissue culture offer several benefits. They are free from pesticides and unwanted extraneous organisms such as parasites, pathogens, snails, planarians, insect larvae, algae and annoying "weeds" such as duckweed. The pesky removal of rock wool is omitted completely and with one portion we get quite a large number of small individual plants.

With the increased interest of home owners in incorporating ornamental aquatic plants in their landscape, and with the large number of aquatic plants that have unique flowers and leaf shapes, now is the time to consider establishing a culture facility to meet the demand of ornamental plants for the country.

5

Feed Management

Feed plays a vital role not only in growth but it also enhances the market value of ornamental fish. In natural waters, fish has access to a variety of natural food items, but their overall wellbeing depends upon the availability of their favorite food. However, in a confined system, nutritionally balanced supplementary feed is indispensable due to limitation on availability of natural live fish food organisms. In general, ornamental fish are fed with a variety of commercially available pelleted feeds, imported from countries like Singapore, Hong Kong, Korea and Thailand etc. Although they are very popular among ornamental fish hobbyists, the high cost of these feeds is not economical for commercial-scale application at farmer's level. Further, in contrast to food fish in a pond, feeding ornamental fish in smaller culture units need high precision. Therefore, it is essential for an ornamental fish producer to possess technical know-how regarding nutritional requirements and feeding behaviour of different species in order to formulate farm-made feeds by using locally available low-cost ingredients.

Formulation of low cost and nutritionally balanced supplementary feed, using locally available ingredients, is the key to the success of ornamental fish culture in India. To reduce the cost of feed and make the ornamental fish production venture more profitable, the feed can be prepared at farm by using cost-effective and locally available quality ingredients (plant/animal-based).

Identification and selection of the feed ingredients are essential and it should be according to the fish species to be cultured. A number of feed ingredients may be locally available which can be used for feed formulation. Fish culturists have to identify and choose the best ingredients, depending upon its easy availability, nutrient status and cost, keeping in mind the nutritional requirement of ornamental fish (size, feeding habits, etc.) to be cultured.

Types of Feed

The formulated feeds, used for ornamental fish, are mainly of two types depending upon the moisture content i.e. dry feeds and non-dry (moist/wet) feeds.

a) **Dry Feeds:** These are made from dry ingredients and the moisture content varies between 6-10%. The different forms of dry feed are:

Mash meal: A simple mixture of dry ingredients which can be used for small sized fish larvae (fry).

Pellets: The dry feed, compacted into a definite shape by mechanical means, is termed as pellets. A hand-operated or electric pellet making machine can be used at small farms, whereas large farms can set up a feed mill.

b) **Non-dry Feeds:** These are of two types.

Moist Feeds: These are mixtures of either both wet and dry ingredients or only dry ingredients with added moisture. The moisture content of moist feeds varies between 18 to 40%.

Wet or Paste Feeds: The wet feeds are made from wet feed ingredients and fed through mesh net or sieved platform. It generally includes wet ingredients such as trash fish, shrimps, beef heart etc. or live food with 45-70% moisture. The wet feeds are mainly used for feeding the young ones, carnivorous species and brooders.

Formulation of different types of feeds

The feed is formulated according to the growth stage & size of a fish.

a) Larvae/fry: Mash meal, wet/paste feed/live food

b) Fingerlings/ grown-up/brooders: Pellet feed/live food

Ingredients are selected and feed is formulated with respect to the feeding habits and size of a fish.

a) Herbivorous & omnivorous: Dry/moist (plant-based ingredients / live food)

b) Carnivorous: Dry feed (animal-based ingredient like a fish meal) /moist feed (trash fish, shrimps, beef heart) / live food (tubifex, earthworms, insect larvae, blood worms, etc.).

Storage of feed

For reasons of cost and convenience, dry diets are presently the most widely used feeds in aquarium. These include extruded feeds, hard pellets, crumbles, and flakes. The general rule for preservation of these feeds is to store them in a dry, well-ventilated area that affords some protection from rapid changes in temperature. Cooler temperatures are the best, although actual ambient temperature is less important than minimizing extreme changes. A good storage facility should also provide adequate containment for control of pests. No matter what type of feed is used, there is little or nothing that can be done to enhance its potential storage stability once it has been delivered to the farm. A practical knowledge of the most important factors that contribute to feed degradation and a little attention in maintaining proper storage conditions can significantly minimize loss of vitamin potency, mold growth, fat rancidity and infestation by insects and rodents.

A proper storage facility should be developed for feed storage to prevent spoilage and maintain the keeping quality. Proper storage is very essential to maintain the nutritional quality of feed. The storage facilities of both dry and wet feeds are different. A feed has to be stored properly under hygienic conditions to avoid any kind of infestation and spoilage. The dry feeds are less vulnerable to spoilage due to low moisture content (about 10%), hence, are easy to store at room temperature in a moisture-free environment for a longer period of time. The moist feeds are highly vulnerable to fungal, bacterial and parasitic infestation due to their high moisture content if not stored properly at low temperature. A fungus infested rotten wet feed with high bacterial load may cause diseases and even fish mortality. Therefore, the wet feeds need to be used fresh or be consumed within the shortest possible time after preparation to prevent spoilage. The moist feed should be stored under refrigeration only for a short period. Dry mash/meal or pelleted feed should be stored for 2-3 months in a ventilated moisture-free environment. Moist feed and wet feed can be stored at low temperature (-20°C), for one week, but it is better to use in fresh condition.

Feeding mechanism

A right type of feed dispensing method is mandatory to minimize the wastage and to prevent deterioration of water quality. Feeding method should ensure a sufficient quantity of feed to the whole stock. Hand/tray feeding can serve all the fish in small size ponds/tanks/cisterns/pits. However, an automatic or demand feeder is an efficient alternative to feed the fish more effectively, without much wastage in large size culture systems.

Hand feeding: A required quantity of feed is dispensed at fixed place and time, preferably by the same person every day. In this type of practice, the quantity of feed provided to fish should be pre-decided to avoid wastage due to overfeeding. Fish should not be fed too frequently or too much to prevent feed wastage and pollution.

Tray feeding: Trays are placed in different places in the pond. Feed (dry/dough/non-dry) is provided in meshed/plastic trays at different places in the culture ponds/tanks. Tray feeding gives correct information regarding the utilization of feed by the cultured fish, which helps us to adjust the feed quantity in a more precise way.

Automatic demand feeders: The automatic demand feeders are used to dispense the calculated amount of feed depending upon the fish stock in the pond. The fish can take the feed according to the need. The Automatic demand feeders are effective and time-saving devices to dispense pelleted feeds.

Feeding rate, frequency and time

The feed residues, suspended in the column and deposited at the pond bottom, can cause pollution, resulting in an increased risk of high organic load, high BOD and reduced dissolved oxygen, thus resulting in poor growth and increased mortality rate. In order to avoid over or under feeding of the fish, it is very important to work out the correct feeding rate at an appropriate time of the day. Feeding rate, time and frequency depend on the stage as well as the body weight of the fish. Further, acceptance and utilization of feed also depend upon the optimum environmental conditions like temperature, DO, etc., because metabolic activities of fish are directly related to these conditions.

Estimate correct biomass of fish stock for calculating the right amount of feed. It is very important to keep track of total number, average size and weight of the fish in the tank. The amount of feed required per ration is calculated as follows:

Total amount of feed = Average fish size (body weight) x Feed rate (%) x Total number of fish in the pond/100

We should avoid both overfeeding and/or underfeeding. The use of supplementary feed is very essential to obtain high production and good returns. But, it increases the cost of production so it should be used judiciously. The overfeeding increases the cost of inputs and also deteriorates the water quality. It is very commonly stated that a fish may not die due to underfeeding but overfeeding. However, underfeeding is also not advisable as it leads to poor growth of fish.

Food conversion efficiency of feed

It is important to know that how much of given feed is being converted into fish biomass. It is represented in the form of feed conversion ratio (FCR). It gives information regarding the total feed consumption and its output in the form of fish biomass. FCR can be calculated with the following formula :

FCR = Total feed given (kg)/Total fish biomass produced (kg)

It is advisable to select the right time for feeding. Feed intake and digestion capacity of fish depend upon various environmental factors like temperature, pH, DO, etc. Factors like very high/low temperature or DO <5.0 mg/L affect the fish metabolism directly. Hence, it is important to provide feed at the time when fish can consume it and stay stress-free during the digestion process. Start feeding fish after sunrise at a fixed time daily and never feed during late evening or night hours.

Selection of correct feeding frequency is also required. Fish grows fast during initial days and need to be fed at frequent intervals to support their metabolic activity and overall growth. However, the feeding frequency can be reduced or increased depending upon various factors including growth rate, water quality parameters, environmental conditions, etc. Fish culture tanks are not a dumping pit.

Feeding fishes of one week to three weeks age

- Feed should not be put in the tank till they are free-swimming because for the first few days they feed on their yolk sac.
- Newly hatched brine shrimp is the best food for the tiny angel fry. Frozen brine shrimp is one of the decent substitutes for the live one.
- Brine shrimp are fed live to the young ones at first to make sure that no excess food is floating around in the tank.
- Feed the fish only at the time when they need.
- Once the fry is full and has little orange bellies about the size of a pinhead, it's safest to siphon out the uneaten brine shrimp, if any.
- Feeding is recommended for 3-4 times per day.
- Feeding in light quantities decreases overfeeding and associated problems such as ammonia and diseases.

Feeding during larval rearing

- When the eggs hatch out, the larvae that emerge have a large, yellow sac and are barely able to swim.
- The larva feeds on the yolk sac, until and unless all the yolk sac is absorbed, the fry does not start feeding on external food.
- Young ones of egg layer fish should be very carefully raised from the moment they hatch.
- Sudden fluctuations in the water temperature should not be allowed to prevail as the young fish cannot tolerate these changes.

- Consistent water exchange and good feeding are key to the fast growth of the fry.
- The fry attains a size in which they will accept finely crushed flake foods.
- As a food supplement, flake foods are provided in small quantity.
- After three weeks, brine shrimp can be fed and must be eaten by fish within 15 minutes of adding it to the aquarium.
- At the age of five weeks, the young fish are introduced to dry feeds.
- Until the seventh week, a small amount is fed twice daily.
- During this time, the small fish attempts to eat the dry flakes but they usually spit it out soon after taking it into their mouths.
- Some fishes may eat the flakes but some will not.
- Around the seventh week, the fish begin accepting dry flakes and there should be few flakes, if any, remaining on the bottom of the tank like the previous weeks.
- At about 6 weeks of the age, the young fish reach a size which begins accepting blended beef heart cubes.
- Newly hatched tiny brine shrimp can still be given to the young fish up to 3 months but beef liver and flakes are all that is necessary for quick growth of the young ones.

Live fish feed and their culture

Live organisms are the natural fish food organisms essential for the growth and quality of ornamental fish, as they contain all the essential nutrients (proteins, carbohydrates and fats) including micronutrients (vitamins and minerals). Use of live feeds enhances the survival, growth and breeding efficiency of the fish besides providing pigments for colour development. Live food is available in abundance in many types of water bodies but it is very difficult to collect. Secondly, the quality of food available in natural conditions is uncertain and it could also be a potential source of disease transmission. Therefore, live food culture unit needs to be incorporated as an integral part of the ornamental fish production unit.

Classification of natural live fish food

Live foods are categorized as follows:

1. Plankton: a) Phytoplankton b) Zooplankton c) Bacterioplankton
2. Periphyton

3. Macrophytes
4. Benthos (Zoobenthos)
5. Miscellaneous

Among the live foods mentioned above, zooplankton, some zoobenthos and miscellaneous live foods are mainly used for ornamental fish feeding.

1. Plankton

- Microscopic, passively or weakly free-floating organisms.
- Divided into three types, namely, a) Phytoplankton and b) Zooplankton c) Bacterioplankton.
- On the basis of their size, they have been classified as:
  - i) Ultraplankton (0.5-10 μm)
  - ii) Nanoplankton (10-50 μm)
  - iii) Micro or netplankton (50-500 μm)
  - iv) Macroplankton (>500 μm)

A. Phytoplankton

- Plankters of plant origin,
- Mainly consists of chlorophyll (a, b as green, c as golden-green and d as red pigments, besides carotenoid pigments xanthophylls, phycoerythrin etc.) bearing unicellular microscopic autotrophic algae.
- Planktonic algae are divided into eleven classes mainly based on their pigmentation and food storage and among these four major classes namely:

  i) Chlorophyceae ii) Bacillariophyceae iii) Cyanophyceae or Mixophyceae and iv) Euglenophyceae are important as live foods.
- Chlorophyceae and Bacillariophyceae are more desirable algae for aquaculture.
- Myxophyceae and Euglenophyceae always form the bloom, which is detrimental to aquaculture. Besides, most of the algae of Myxophyceae produce toxins, which cause health problems in fish.

i) Chlorophyceae (Green Algae)

- Green pigmented algae with chlorophyll a, b gives green colour to the cell and are of the following types:

- Unicellular free-floating non-colony-forming, examples, *Chlamydomonas, Chlorella, Tetraselmis, Isochrysis, Closterium*;
- Unicellular colony-forming, examples, *Volvox, Pandorina, Eudorina*;
- Multicellular unbranched filamentous, examples, *Spirogyra, Ulothrix, Oedogonium, Zygnema;*
- Multicellular nonfilamentous, examples, *Pediastrum, Scenedesmus, Tetradesmus, Cosmerium, Actinastrum.*

ii) *Bacillariophyceae (Diatoms)*

- Commonly known as diatoms.
- Golden brown colour due to chlorophyll c and carotenoid pigments ß-carotene and xanthophylls.
- Diatoms are two types namely :
  i. Centric diatoms, examples- *Coscinodiscus, Stephanodiscus, Cyclotella*;
  ii. Pennate diatoms, examples- *Navicula, Chaetoceros, Skeletonema, Pinnularia, Synedra, Fragilaria, Nitzschia*.
  iii. Centric diatoms are round solitary or colonial and radially symmetrical, whereas pennate diatoms are elongated and bilaterally symmetrical.

iii) *Cyanophyceae or Myxophyceae (Blue-Green Algae)*

- Commonly called blue-green algale.
- Major pigments are chlorophyll a, b, and carotenes, xanthophylls and phycobilins (phycocyanin and phycoerythrin).
- Unicellular or multicellular thalloid or filamentous, colonial or noncolonial forms and usually with a gelatinous sheath.
- Some common blue green algae are *Microcystis, Aphanocapsa, Tetrapedia, Gleocapsa, Nostoc, Anabaena, Oscillatoria, Spirulina, Gloeotrichia, Calothrix*. These plankton are not used for ornamental fish feeding.

iv) *Euglenophyceae (Euglenoids)*

- Green in colour due to presence of chlorophyll a, b and ß-carotene and xanthophylls.
- Unicellular flagellates having 1 to 3 flagella, originating anteriorly from an invasion; mostly free swimming.

- Common euglenoids are *Euglena, Phacus* and *Trachelomonas*.

Among all these phytoplankton, only *Chlorella* is important for ornamental fish feeding and zooplankton production. As having good nutritive value, *Spirulina can* be used as one of the ingredients of artificial feed.

B. Zooplankton

- Microscopic free-swimming animal components of aquatic systems are represented by a wide array of taxonomic groups; namely: Protozoa, Rotifera, planktonic forms of Crustacea including Cladocera, Ostracoda and Copepoda.
- Their larvae are the most common food for ornamental fish culture and breeding.
- They serve as "Living Capsules of Nutrition".

Providing appropriate live food organism at an appropriate time plays a major role in achieving optimum growth, survival and breeding performances in ornamental fish.

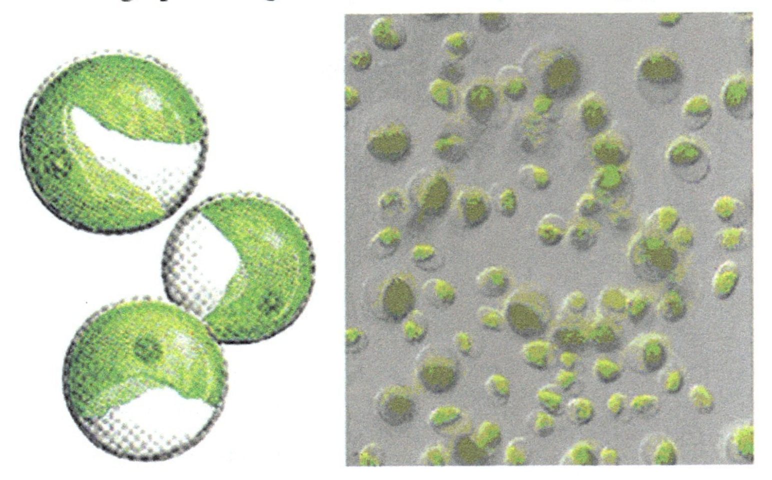

Fig. 43: Chlorella

i) Protozoan

- This group does not directly contribute much to the natural fish food organisms, but indirectly they are involved in the basic fish-food cycle.
- Common examples are infusoria (*Paramecium* and *Stylonychia*), *Arcella, Difflugia, Actinophrys* and *Vorticella*.

- Infusoria are the most primitive of all organisms in the animal kingdom, belonging to the class Ciliata under the phylum Protozoa.
- Besides being small in size, they are soft-bodied and nutritionally rich.
- Owing to these qualities, they serve ideal live food for ornamental fish.

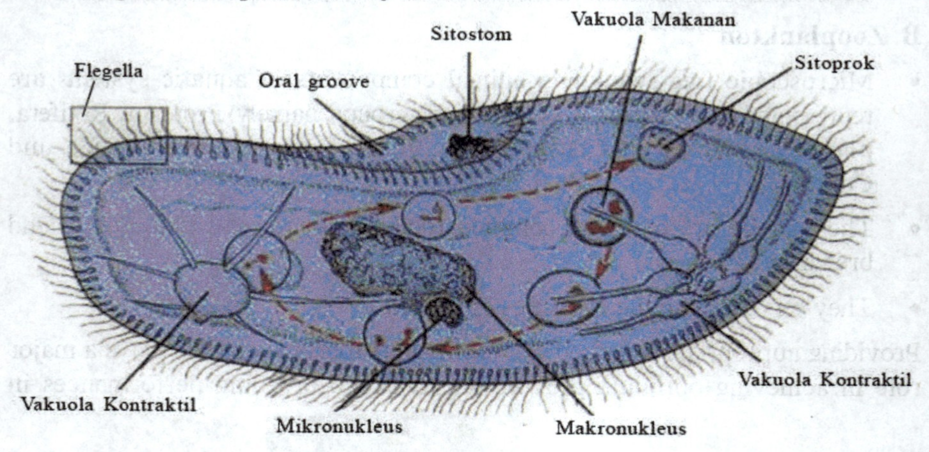

Fig. 44: Stylonychia

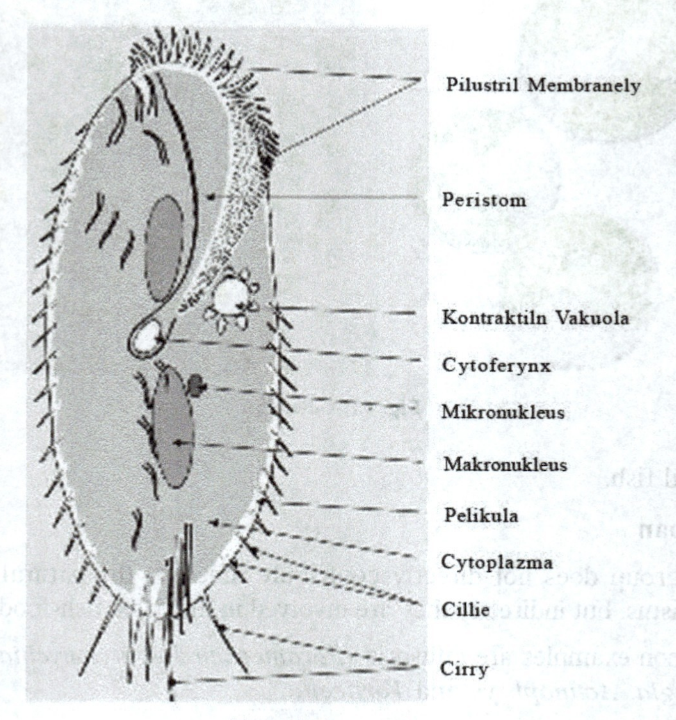

Fig. 45: Paramecium

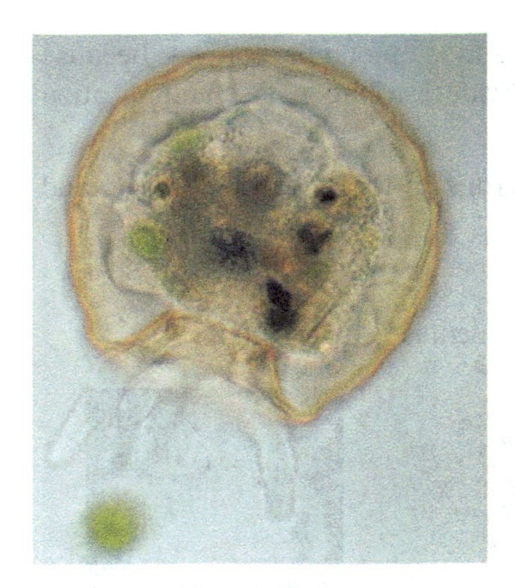

Fig. 46: Arcella

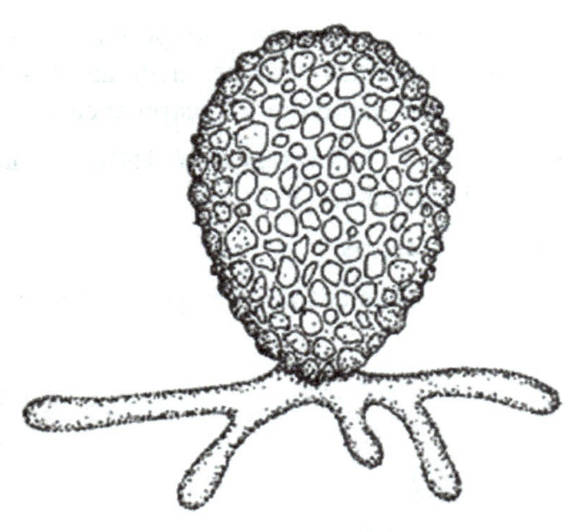

Fig. 47: Difflugia

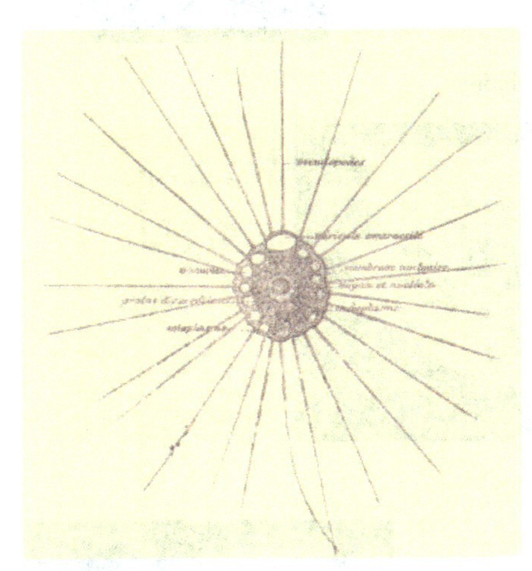

Fig. 48: Actinopyris

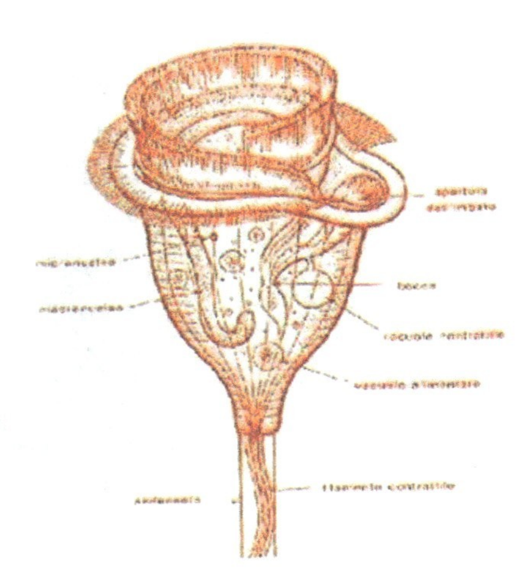

Fig. 49: Vortecilla

- The commonly used freshwater infusoria are *Paramecium* and *Stylonychia.*

ii) Rotifera

- Commonly called as wheel animalcules.
- Slow-moving rotifers are the most important group of microscopic animals for culture and seed production of ornamental fishes.
- Some of the common species of rotifers occurring in tropical waters are *Brachionus, Keratella, Asplanchna, Polyarthra, Kellicotia* and *Filinia.*

- *Brachionus* spp. are more popular as live food among the rotifers, because of its high nutritive value, small size, worldwide distribution, fast multiplication and easy adaptability to captive culture.
- The size of it varies from 150-250 microns in length (without foot) and 100-150 micron in width.

iii) Crustaceans

Following crustaceans are important for ornamental fish nutrition.

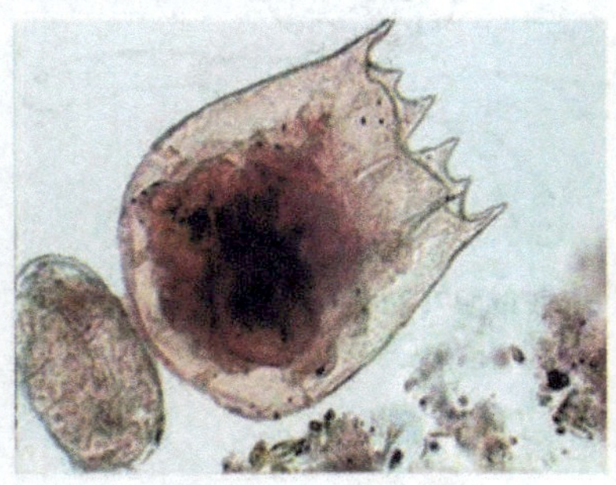
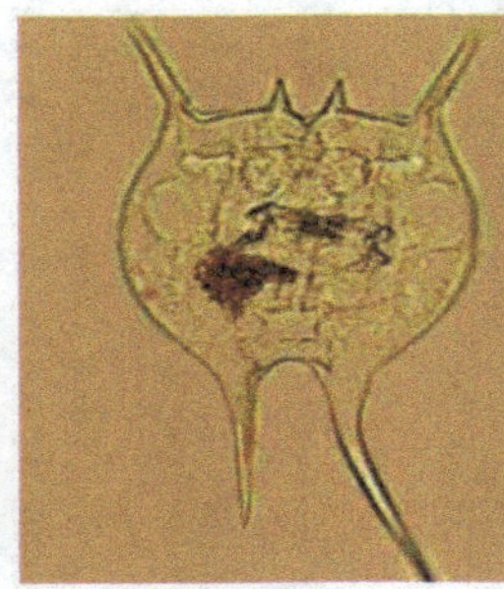
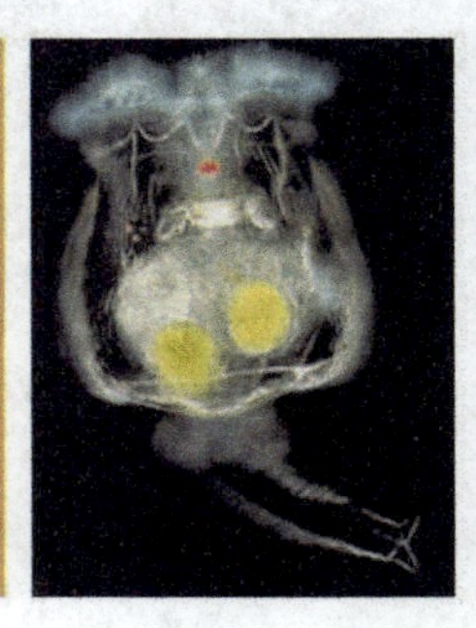

Fig. 50: Branchionus

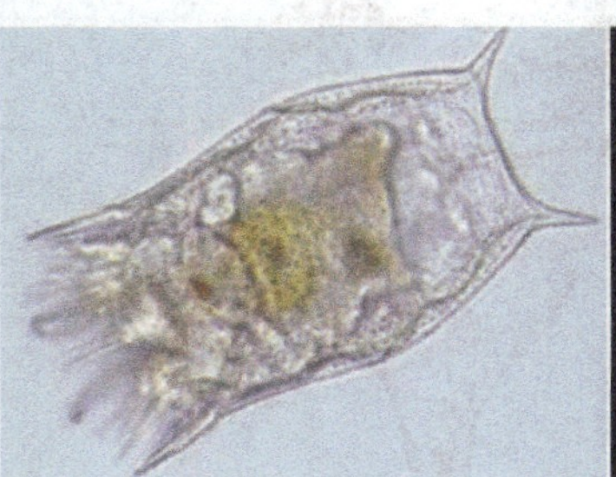
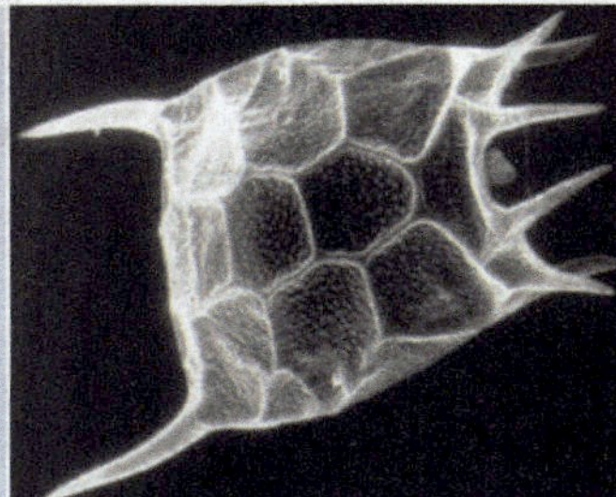

Fig. 51: Keratella

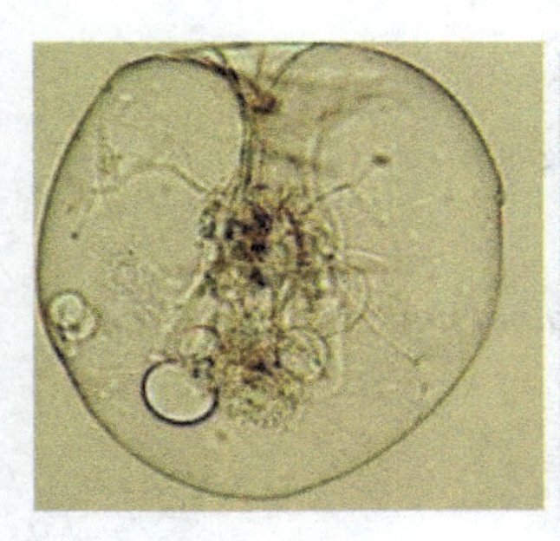
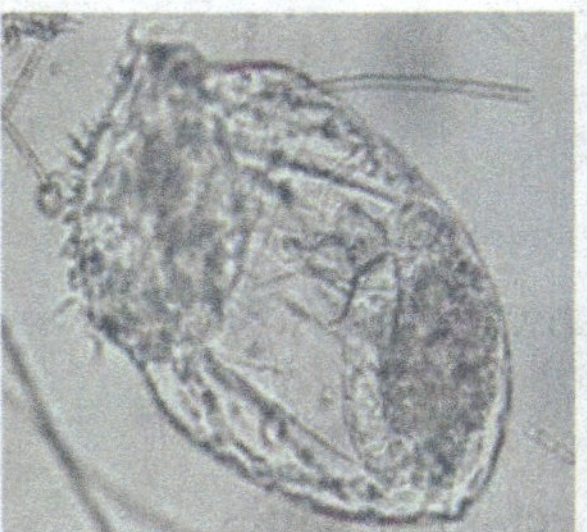
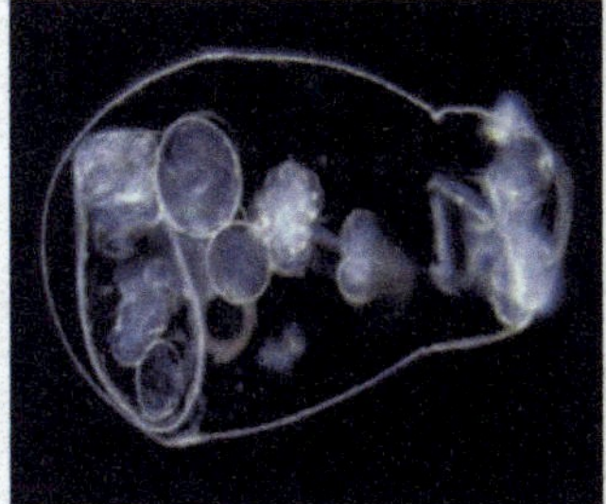

Fig. 52: Asplanchna

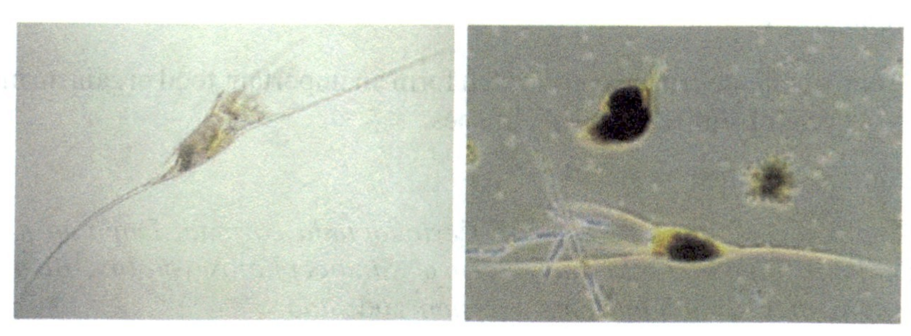

Fig. 53: Kellicotia

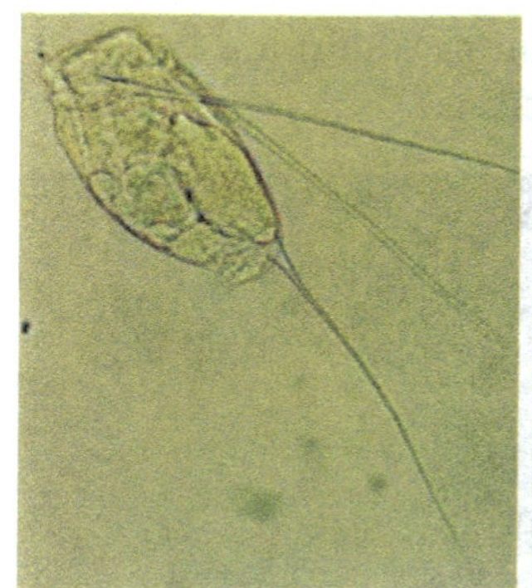

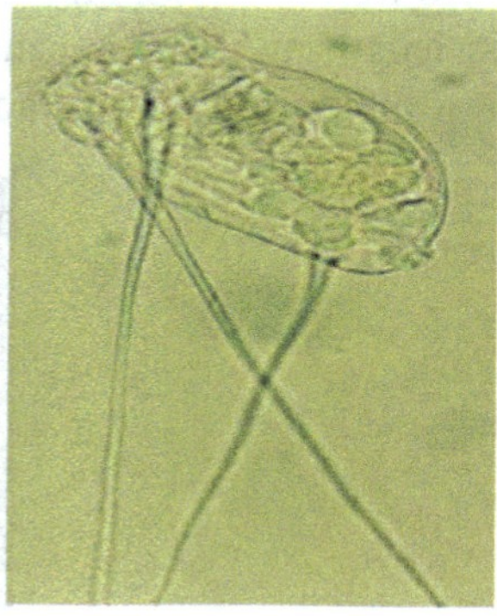

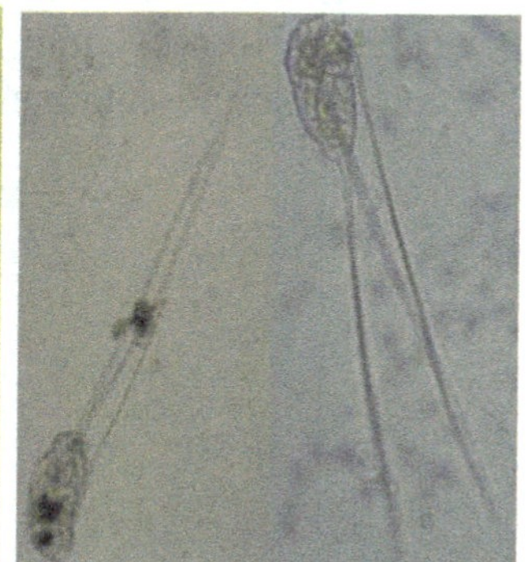

Fig. 54: Filinia

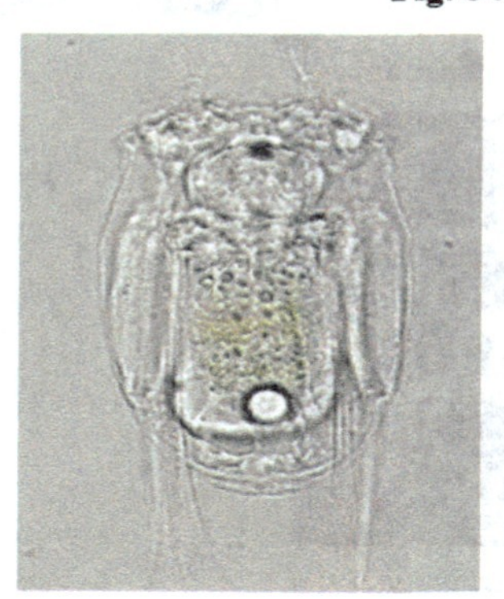

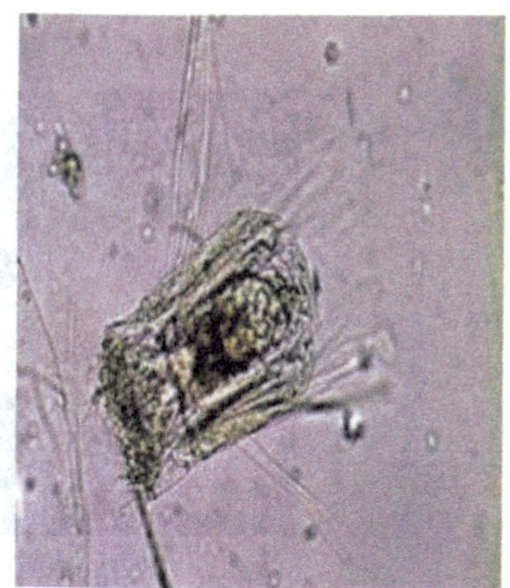

Fig. 55: Polyarthra

a) Cladocerans

- Group of minute crustaceans, which form an important food organism in the food cycle of finfishes and shellfishes.
- Commonly called "water fleas".
- Common cladoceran species are *Ceriodaphnia cornuta*, *Daphnia pulex*, *Daphnia lumholtzi*, *Moina micrura*, *Simocephalus vetulus*, *Bosmina longirostris*, *Diaphanosoma excisum* and *Sida*.
- Among this *Moina* (0.5-1.0 mm in length and 0.2 –0.6 mm in width) and *Daphnia* (0.5-2.5 mm in length and 0.3–1.0 mm in width) are the most common representatives of this group of crustaceans.

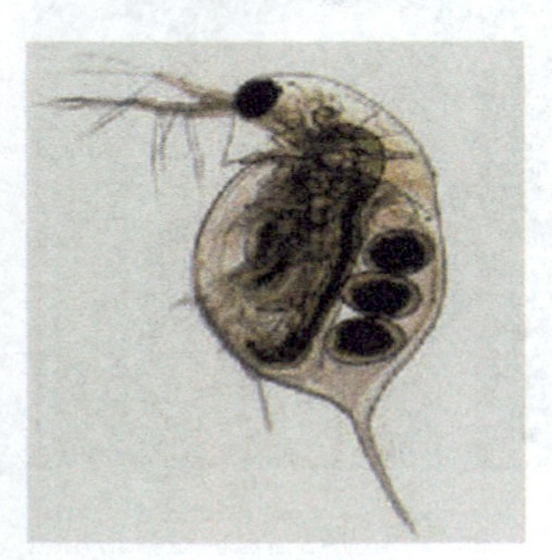

Fig. 56: Cereodaphnia

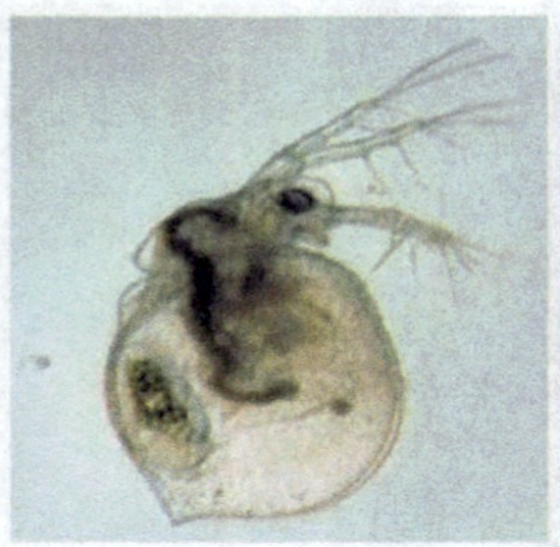
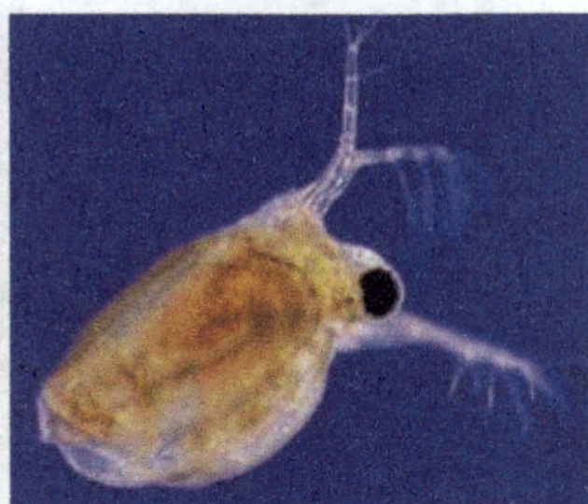

Fig. 57: Daphnia

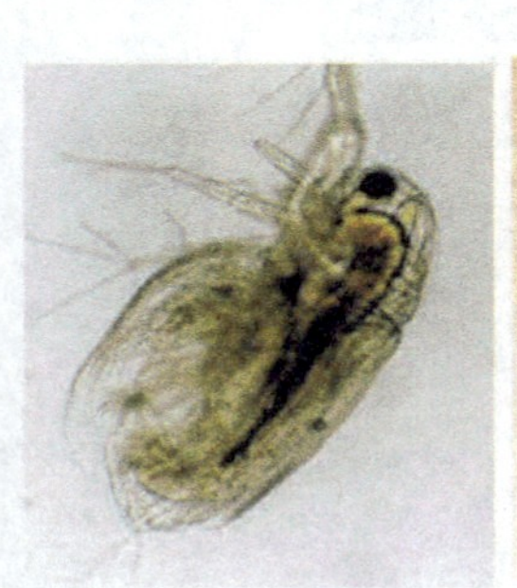
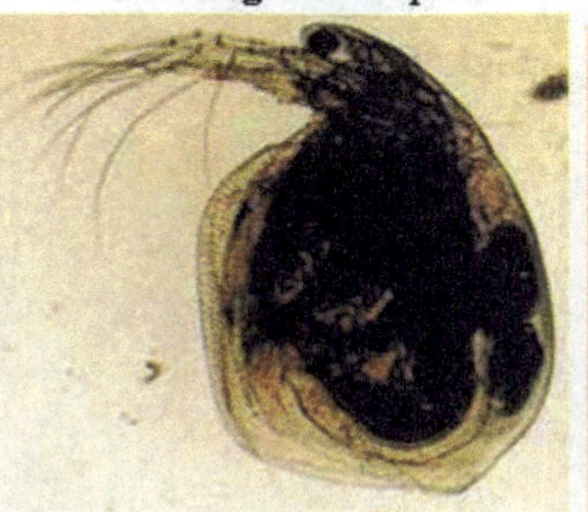
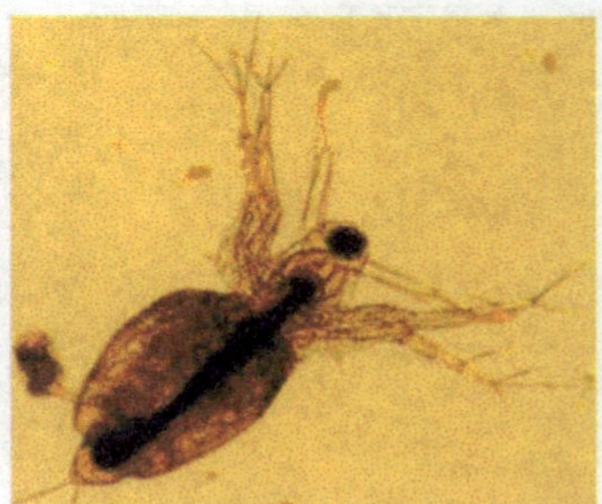

Fig. 58: Moina

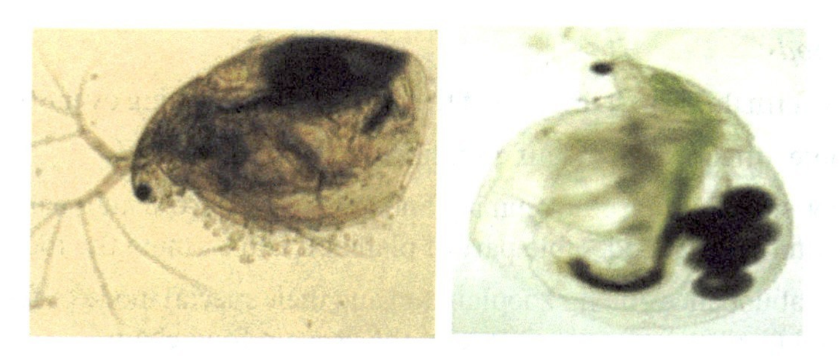

Fig. 59: Semocephalus

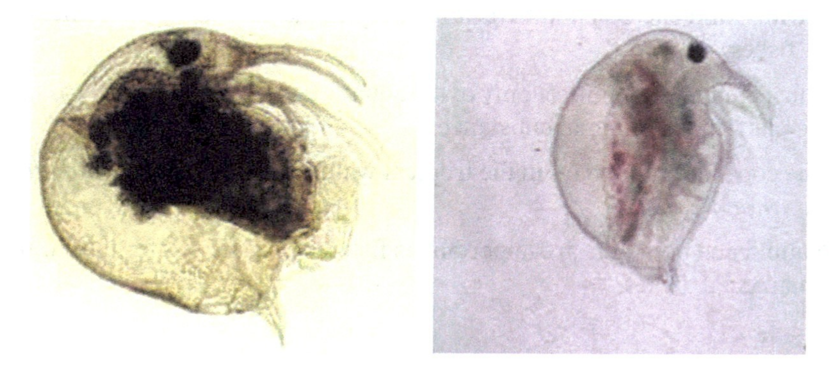

Fig. 60: Bosminia

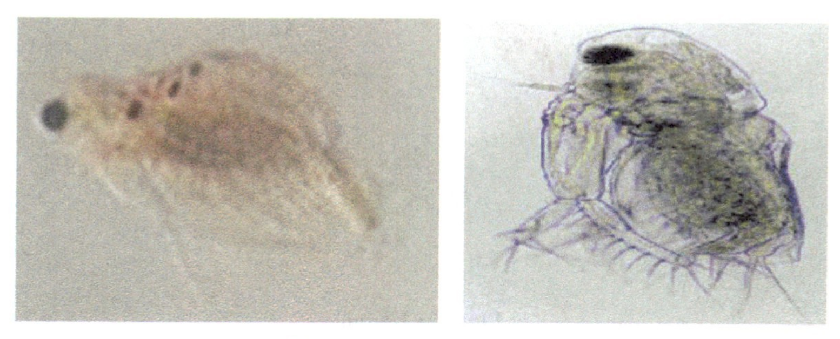

Fig. 61: Diaphanosoma

Fig. 62: Sida

b) Copepods

- They form the largest division of the crustacea comprising over 6000 species.
- Minute, often less than half a millimetre in length.
- Major species are free swimming and abundant in fresh and marine habitat, constituting a considerable part of plankton and forming the food of fishes.
- Their abundance among zooplankton and their special mode of feeding have earned them an important place in the aquatic food chain.
- They graze directly on the primary sources of energy, namely, phytoplankton, detritus, and bacteria and in turn fall prey to the secondary consumers such as fishes.
- Thus, their role in the economy of the aquatic ecosystem is mainly to convert the plant into an animal substance.
- The copepods, abundant in the tropical water, are *Cyclops, Diaptomus* and *Canthocamptus*.
- Nauplii and preadults are important as fish food organisms for the culture of ornamental fishes.

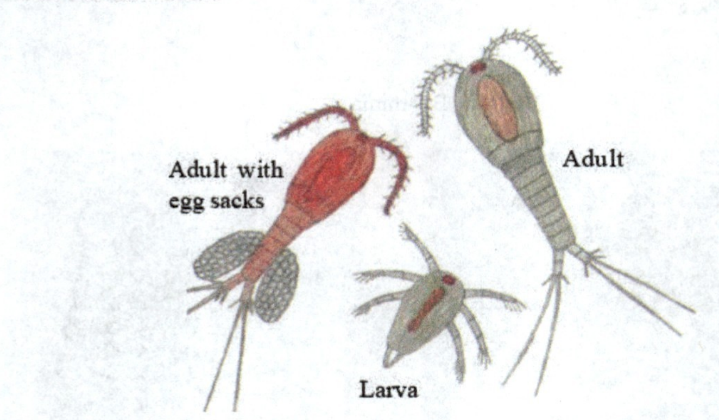

Fig. 63: Nauplii

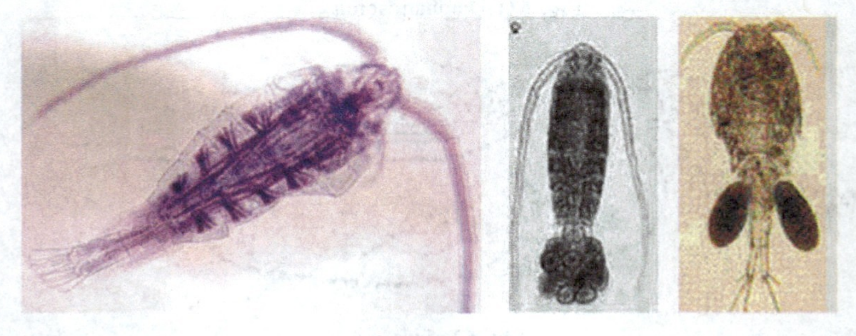

Fig. 64: Cyclops

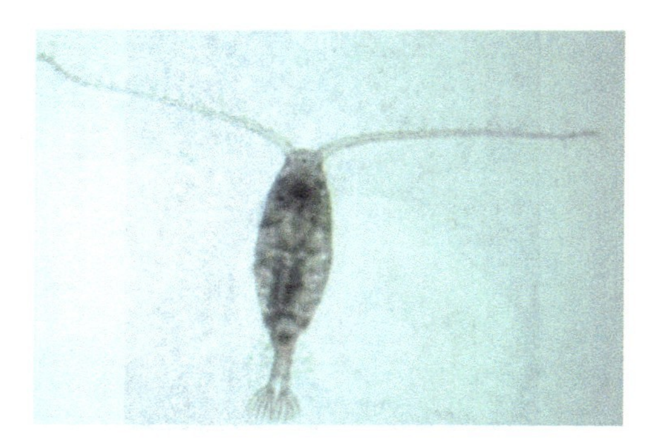

Fig. 65: Diaptomus

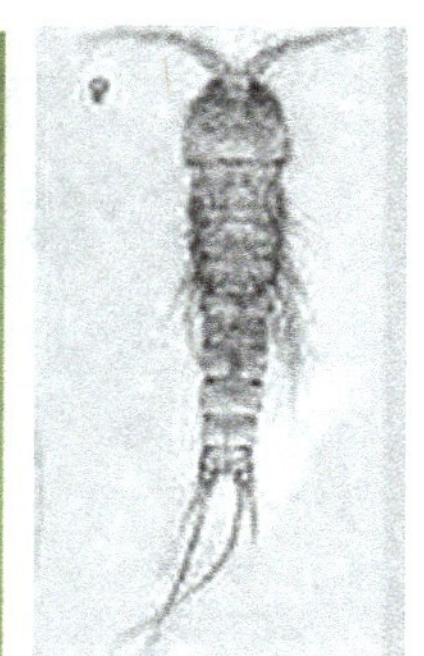

Fig. 66: Canthocamptus

c) Ostracods

- Small, bivalved crustaceans with a laterally compressed body, not distinctly segmented and found in fresh as well as in marine waters.
- There are about 1700 species of known ostracods in which about 35% inhabit freshwater habitats.
- They are usually called "seed shrimps".
- As the name implies, the body is enclosed within a bivalve shell which is hinged dorsally and encloses the entire body within the shell.
- A group of adductor muscles control the movement of the valves.
- These animals are oval or bean-shaped and found in pools, streams especially in shallow areas where aquatic weeds or algae are abundant.
- Although in captive culture system they do not form an important food, in the natural aquatic habitats, they are often included in the fish food cycle.
- The common examples are *Cypris* and *Stenocypris*.

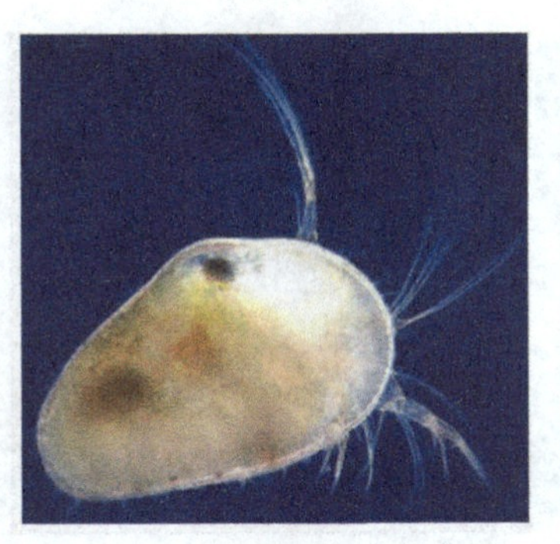

Fig. 67: Cyprus

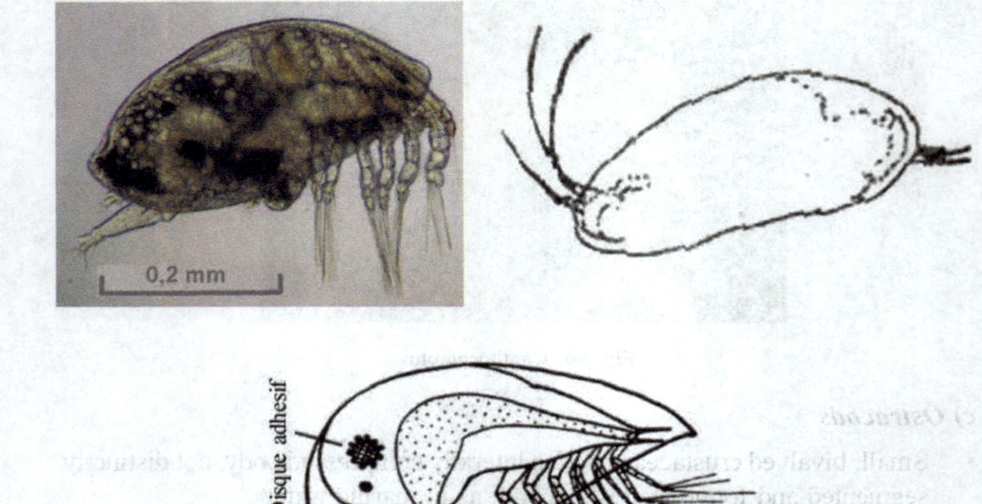

Fig. 68: Stenocypris

d) Anostraca

- The most common anostracans is Brine shrimp or *Artemia salina*.
- It is one of the best live foods.
- It is also called Brine worm or Sea monkey.
- It is a minute crustacean and can be collected from saltwater ponds around the world.

- *Artemia nauplii* contain all essential fatty acids and significant concentration of vitamins and carotenoids in addition to the protein content of 50-60% depending upon the strain and most accepted live feed for ornamental fish.

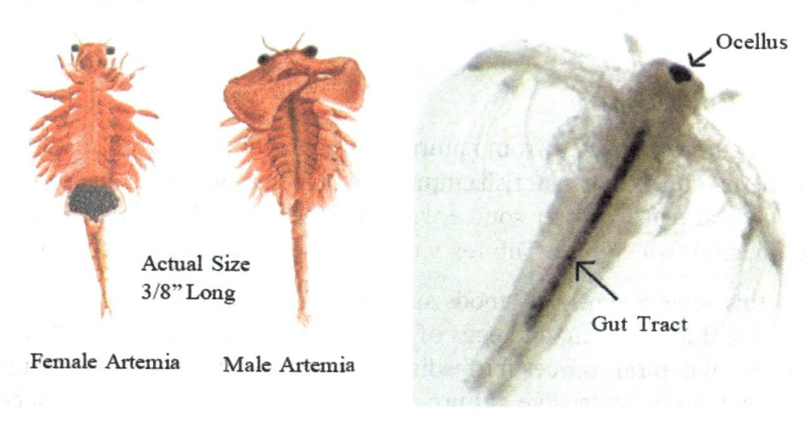

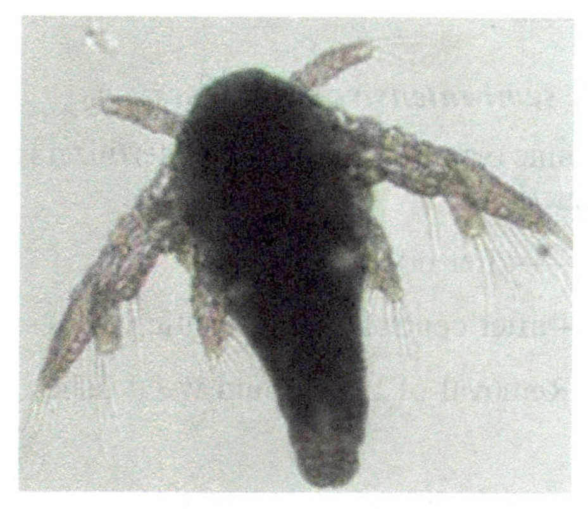

Fig. 69: Adult artemia and its nauplii

C. Bacterioplankton

- Free-floating bacteria of the aquatic environment are the bacterioplankton.
- These are not important live food for ornamental fish in captive condition.

2. Periphyton

- All microscopic organisms (both plants mainly algae and animals mainly microbes) which grow attached on materials submerged in water are known as periphyton.
- These are not important live food for ornamental fish in captive condition.

Sources of Live foods

There are two sources of live foods. i) Natural collection ii) Production in the semi-intensive culture system and iii) Mass culture of live foods under controlled conditions.

i) Natural Collection

Planktons can be collected from natural sources with the help of plankton net for hatcheries and ornamental fish culture and breeding. Artemia can be collected from a natural marine water source with the help of hand net and can be used for ornamental fish culture. Tubifex worm can be collected from sewage drain.

Besides this several other live foods such as different zoobenthos, earthworm, white worm, micro worm, and eggs of ants can be collected from both aquatic and terrestrial natural sources for feeding of larvae in hatcheries and ornamental fishes. For the semi-intensive culture of fishes, live foods can be produced in the same culture system.

ii) Production in the semi-intensive culture system

A combination of organic manures and inorganic fertilizers is preferred in ponds for natural food production.

Seasonal pond————after drying

Perennial pond———after controlling of aquatic weeds,

Removal of Insects and weed fishes

↓

Inorganic fertilizers, limestone is applied based on the soil pH

After 7 days manuring and fertilizing

iii) Mass culture of live foods

For using the live food in hatcheries and for ornamental fish culture and breeding, this can be cultured under the controlled conditions for its regular availability. Culture process of some important live foods is discussed below:

a) Culture of Phytoplankton

- Among phytoplanktons, some are filamentous and some are non-filamentous.
- Non-filamentous phytoplanktons are desirable live foods for ornamental fish.

Among non-filamentous phytoplankton, *Chlorella* is well accepted live food for feeding of larvae in hatcheries.

Mass Culture of Chlorella

Chlorella culture pool or tank is filled up with water.

Enrichment with Yashima Medium (Ammonium sulphate-100 g/1000 L, Single superphosphate-10 g/1000 L, Urea-10 g/1000 L) Trace element solution (Manganese chloride, $MnCl_2$. $4H_2O$-1.81 g/L, Molybdenum trioxide-$MoO_3.4H_2O$-0.0177 g/L, Zinc sulphate, $ZnSO_4.4H_2O$-0.222 g/L, Cupric sulphate, $CuSO_4.5H_2O$-0.079 g/L, Boric acid, H_3BO_3-2.86 g/L)1 ml/L).

- After that Chlorella @ 10^2 cells/ml can be added. When density is 10 X 10^6 to 20 X 10^6 cells per ml harvesting should be done with the help of plankton net.

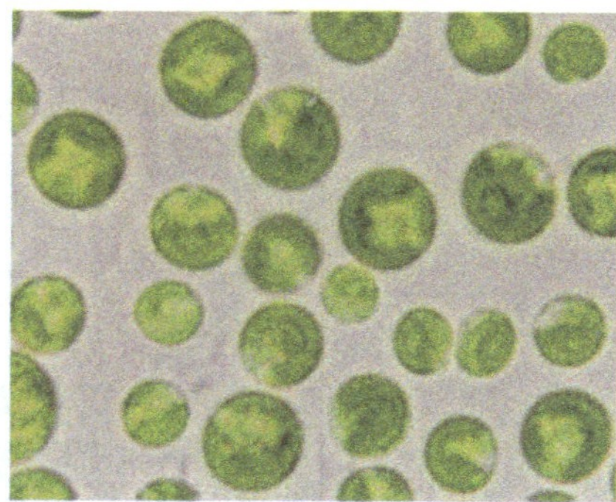

Fig. 70: Chlorella Culture

b) Culture of Zooplankton

Mass Culture of Protozoa

- *By using banana peelings*
- *By using milk*
- *By using hay or straw infusion*
- *By using lettuce leaves*
- *By using apple snail*
- *Banana peelings method*

Collection of banana peelings

↓

Place in a clean big jar (50L)

↓

Add filtered water

↓

Keep in a cool place

↓

Cover with mosquito net cloth

↓

Aquarium set-up with banana peelings for the culture of infusoria

↓

The culture should be kept undisturbed for 2 days

↓

The water turns yellow, smells foul due to decomposition of banana peelings by bacteria

↓

A film of slime will be formed on the water surface

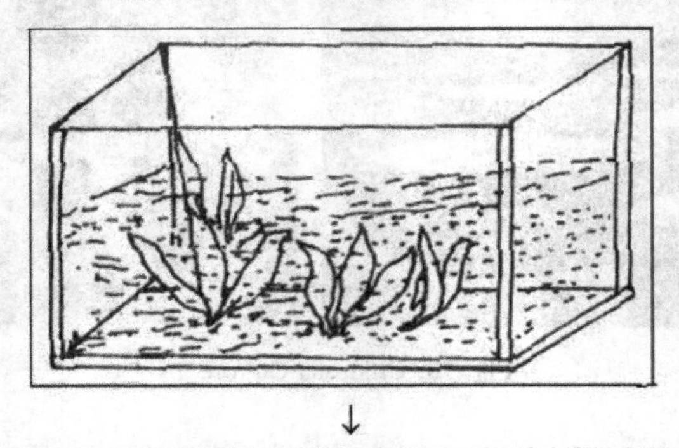

↓

After 4-5 days, the water turns clear with transparent light yellowish colour, because of the floating spores of infusoria which feed on bacteria and multiply in large numbers

↓

When slime on the water surface breaks up and disintegrates, harvest Infusoria

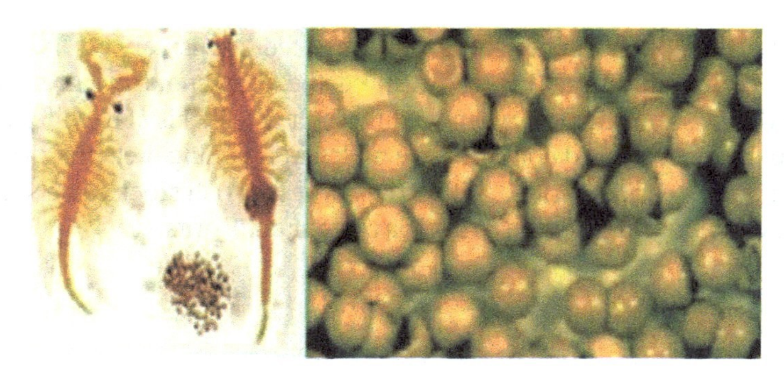

Fig. 71: Collection of artemia cysts.

Mass Culture of Anostracans *(Artemia nauplii)*

- Culture of *Artemia* up to adult is a very cumbersome job and possible only in coastal areas.
- *Artemia* cyst are commercially available, can be collected and the cyst can be hatched out under controlled conditions for the production of *Artemia* nauplii

Color enhancement of ornamental fish through the feed

The pigmentation pattern and intensity of body colours determine the commercial value of an ornamental fish. The de novo synthesis of carotenoids does not occur in fish like other animals but they can convert one carotenoid to others. An ornamental fish being reared in the natural environment easily get colouration due to the availability of plenty of natural food in the form of phytoplankton and zooplankton but under indoor rearing conditions, fish has to be fed on carotenoid supplemented diets.

Carotenoids are found in a variety of natural and synthetic sources that results in yellow, orange and red pigmentation of fish skin.

a) Sources of natural carotenoids

i) Animal Origin: Zooplankton, tubifex, chironomid larvae, Artemia, crayfish meal, shrimp meal, crab meal, yeast, etc.

ii) Plant Origin: Several types of flowers and vegetables

b) Synthetic carotenoids: Astaxanthin, β-carotene, lutein, zeaxanthin, etc. are few of the commercially available carotenoids.

The carotenoids can be easily supplemented in all those formulated feeds which are prepared at the farm. Plant materials, which are to be used as a carotenoid source, need to be grated, dried under shade, grounded and mixed with the other ingredients before pelletization. Fish are fed intensively on live food for the first one month automatically fulfilling the carotenoid requirement. The carotenoids are supplemented when feeding with dry feed is started for a period of 2-3 months. Carotenoid could also be added in finishing diets for 1-2 months at the end of the rearing period; so that the harvested fish stock to be sold in the market should have bright coloration for maximum profits.

6

Water Quality Management

Learning about water chemistry is often neglected by most of the aquarium owners. By understanding the concepts of water chemistry, they can considerably improve success rate in rearing healthy fishes. It is well known that the quality of water has a direct impact on the health of the fish. But many aquarium owners do not understand the basic internal chemistry of the water, nor do they understand how to correct it or safely adjust it. Until the basics of water chemistry is understood and some common water maintenance techniques are learnt, it will be difficult to maintain a healthy and safe environment for the fish in tank. The water quality is by far the single most important factor for the healthy fish, and the more is known, the better management is expected.

Fish depends for their basic necessities on the water in which they live. The most characteristic feature of an aquarium system is, therefore, the quality of the water available in tanks/aquarium. This water must be obtained from some source, pre-treated to make it suitable for the fish, delivered to the fish in sufficient quantities and maintained in good condition. The water supplied to an aquarium is not pure, but it contains dissolved and particulate materials, some are necessary for up keep of the fish and rests are harmful. Contamination may occur not only at source or from the animals, but often takes place within the aquarium from the materials used for its construction. The number of fish that can be maintained within the aquarium is said to be according to the quality and quantity of the water supplied to an aquarium. However, it is rarely the quantity of water per unit which limits the carrying capacity. The capacity is usually set by the requirement of dissolved oxygen and the accumulation of toxic metabolic products in the aquarium.

Water quality parameters

The most important water quality parameters are discussed below:

pH

Every aquarium owner must have heard about the term pH, but many of them do not know about its importance and the procedure to manage it. Measurement of pH tells about the nature of the water whether it is acidic or alkaline. pH scale ranges from 1 to 14. A pH of 7 is neutral, which means the water is neither acidic nor alkaline. Water having pH lower than 7 is acidic. If the pH is recorded higher than 7, for example 8, the water is alkaline. The pH scale is a logarithmic scale. In a simple language, it means that the pH changes at a ten fold level between each number. For example, a pH of 5 is 10 times more acidic than a pH of 6, and a pH of 4 is 100 times more acidic than a pH of 6. So, if fish are supposed to be at a pH of 7, and the water level is 8, they are in water that is 10 times more alkaline than what they should be. If the pH is 9, then they are in water that is 100 times more alkaline than what they need. So, it is easy to see why even a small change in required pH is stressful and potentially fatal to fish.

These examples really emphasize the importance in matching the fish closely to the expected pH level of water and then closely monitoring the pH. Keeping a fish that requires a pH of 8 with a fish that requires a pH of 6 is just not recommended because one or both will be at a very unacceptable level of pH and will be under a great stress.

There are several different ways to influence water's pH. There are chemical additives that can be added directly to the water that will either raise or lower the pH. More natural agents can be used to alter water pH as well. Peat in the tank or filter will acidify the water. Mineral salts like calcium that are found in limestone or in some shells will cause an increase in alkalinity and pH. There is one important consideration in altering the pH of water and that relates to the mineral content (hardness) of the water. Remember that fish are very sensitive to changes in pH, and rapid changes in pH can cause extreme stress and death. Fish should not be exposed to a change in pH greater than 0.3 in a 24-hour period.

The tap water is usually alkaline and its pH should be tested before adding to the aquarium for water exchanges. It must be adjusted if required. To maintain an aquarium a pH level of 8.2 to 8.4 is required.

Alkalinity

Normal range of alkalinity for aquarium fish is 120-300 mg/l. Low alkalinity can result in a sudden change in the pH level that may be deadly to fish. Water should be changed to maintain proper alkalinity. Total alkalinity is the measurement of all bases in the water and can be thought of as the buffering capacity of water, or its ability to resist change in pH. The most common and important base is carbonate. Total alkalinity is expressed as milligram per liter (mg/L) or parts per million (ppm) of calcium carbonate ($CaCO_3$). In the aquarium industry, total alkalinity may be referred to as "carbonate hardness".

There are many reactions in aquarium that produce acids, and others, like biofiltration, that directly use carbonates. Over time the alkalinity can be "consumed," and if alkalinity is depleted, the pH of the water in an aquarium can plummet, causing extreme stress or death to the fish and adversely affecting biofilter function. Alkalinity can be easily replenished in an aquarium by periodically exchanging a portion of the tank water with new water with a moderate total alkalinity or by adding chemical buffers, such as sodium bicarbonate (baking soda), to the water.

Dissolved oxygen

The aquatic animals get oxygen from the surrounding water to fulfil their requirements. Oxygenation or aeration of the water is of fundamental importance, especially as the oxygen supply is one of the factors which ultimately limit the capacity of a particular volume of water to carry the fish. The dissolved oxygen in water comes either from the atmosphere or green plants. The actual content is the function of temperature, salinity and atmospheric pressure. Low temperature, low salinity and higher atmospheric pressure favour more gas content in the water medium so that more oxygen is available. The continuous aeration is very good husbandry practice since it mixes the water, supplies sufficient oxygen to the fish, removes carbon dioxide and maintains a constant temperature in the tank.

Different types of low-cost air pumps are available in the market, though they are often noisy, are of limited power and some frequently fail to aerate. For one or two tanks, such vibratory diaphragm pumps are acceptable, but a spare pump and several replacement diaphragms should be stocked for regular aeration. The pump should be fixed above the tank level or the air-line fitted with non-return valve to prevent back-siphoning in case the pump is stopped or failed. A loop in the air-line 8 cm (3 inches) vertically above the water level also prevent back-siphoning by absorbing the oscillations when the airflows stops. The air tubes from the pump are connected to the air stones for providing small sized air bubbles that diffuses the oxygen in water. To beautify the tanks it is also connected to the selected toys and also for airlift to under gravel filtration. A micro pore air stone is preferred for diffusion of more oxygen in the water. A dissolved oxygen level of more than 5 mg/L should be maintained in an aquarium.

Nitrogenous waste products

The most adverse changes to the water quality prevails due to worst management practices opted by the aquarium inhabitants themselves. The water quality is impaired by the end products of nitrogen metabolism. These include ammonia (either as the gas NH_3 or ammonium ion, NH_4), urea, uric acid and other nitrogenous substances including proteins and amino acids too. Ammonia is one of the most harmful substances. Higher percentage of un-ionized NH_3 prevails at high pH and temperature. In ammonia poisoning, gills become reddish, body becomes darker in color and fish comes to surface to grasp the air. Acute toxicity level of NH_3 is 0.4 mg/L while, the chronic toxicity level is 0.05 mg/L. This problem is more common in newly set up aquarium when immediately stocked to full capacity. Ammonia can damage the gills at a level as low as 0.25 mg/L.

For immediate removal of ammonia, ammonia detoxifier can be used such as Kordon'sAmquel. The *KordonAmquel* removes ammonia, chloramines, chlorine, and many such organic toxins. One teaspoon of *KordonAmquel* treats ten gallons of water. However, it is the best practice to leave aquarium alone until the bacterial load is sufficient. The bacterial phases do not take place unless the tank is initially stocked with feeder fish which can be removed after treatment. Water should be tested regularly until the ammonia drops to nearly zero. At this time, an increase in the nitrite level is noticed. When the nitrites are removed, it will be safe to stock fish in the tank.

The conversion of the more toxic nitrogenous compounds to less toxic compounds is achieved through organisms residing in water treatment units such as filters. In some aquaria, algae are also used in recycling nitrogen. The process of combating the effects of nitrogenous waste products is facilitated by low stocking density, a high water exchange, aeration of the water, frequent cleaning, and removal of faeces and left over food etc. Special water treatment facilities may also be provided for this process.

Ammonia, nitrite, and nitrate

Ammonia, nitrite and nitrate are the breakdown products of nitrogenous waste in available form in an aquarium. The uneaten food, algae, and bacteria can contribute to the significant waste load in an aquarium. As normal practice of rearing fish, this waste needs to be broken down to eliminate or turn into useful substances that can be utilized by other organisms. There is population of bacteria residing in an aquarium that is responsible for this process. The waste may be broken down in four steps:

- The waste from fish, plants, and food breaks down and releases ammonia.
- This ammonia is very toxic to fish and must be converted by bacteria to nitrite.
- The nitrite is also toxic to fish and must then be converted to nitrate.
- The nitrate is not as toxic and is taken up by plants or algae and used in their growth.

Ammonia is the most toxic product formed in water. The source of ammonia in aquarium water is fish respiration and digestion as well as decaying foods. Freshwater fish begin to be stressed at levels of 0.50 mg/L (parts per million). Ammonia levels should be less than 0.05 mg/L in an aquarium.

Nitrate, nitrite, and ammonia are also removed through the weekly water changes. Because high levels of ammonia and nitrite are lethal to fish, it is critical that these products be efficiently removed or converted to nitrate.

Maintaining adequate population of bacteria is very important which can convert ammonia and nitrite into nitrate, an important part of the water chemistry, and the process is known as nitrification which is expedited through biological filtration. Biological filtration occurs naturally in most of the tanks that have been running for a couple of months. This can be achieved through artificial biological filters, available in the market. The better filters often contain a special area or wheel that is made specifically for providing an optimal habitat for growing these bacteria. While the bacteria will live in a traditional filter and on rocks etc. in the aquarium, the new filters harbor a much higher number and can therefore do a better job of removing ammonia and nitrites.

If a fish tank is overcrowded, or the waste level gets too high through overfeeding or dead fish etc., even a properly functioning biological filter can be overwhelmed and toxic conditions can result. Periodical checking of the ammonia and nitrite levels in the tank with a test kit will ensure that the biological filter is working properly. The healthy plants of the tanks will also help in removal of nitrates. As it takes weeks to months to grow a healthy population of bacteria in a tank, it is important that a tank should be allowed to age before fish are stocked. After the tank ages only a few hardy fish should be added to the tank to check the optimum condition in the tank. If the fish survive well, more fish can be slowly added. After a couple of months conditions should be checked to make sure that the biological filter is not overloaded.

Nitrite poisoning follows closely on the heels of elevated ammonia as a major killer of aquarium fish. It half of fish are killed due to ammonia poisoning, the nitrite level rises and puts other fish at risk again. Anytime ammonia levels are elevated, elevated nitrite will soon follow. To avoid nitrite poisoning, test the water while setting up a new tank, when adding new fish to an established tank, when the filter fails due to power or mechanical failure, and while medicating sick fish. Nitrite poisoning is also known as "brown blood disease" because the blood turns brown from an increase of methemoglobin. However, methemoglobin causes a more serious problem than just changing the color of the blood. It renders the blood unable to carry oxygen, and the fish can literally suffocate even though there is ample oxygen present in the water. The optimum level of nitrite should be from 0.0 to 0.5 mg/L.

The water of wells contains high levels of nitrate and contaminated due to fertilizer, agricultural runoff, or sewage. These nitrates are harmful not only to human beings but livestock too. Nitrate can be removed either by reverse osmosis or by applying specialized nitrate removing chemicals. Although many aquarists run their tanks with extreme nitrate levels, the ideal is a maximum of 5 to 10 mg/L. Levels of 20 to 50 mg/L are too high. When nitrate levels rise to completely intolerable levels, fish will become lethargic and may have open sores or red blotches on their skin.

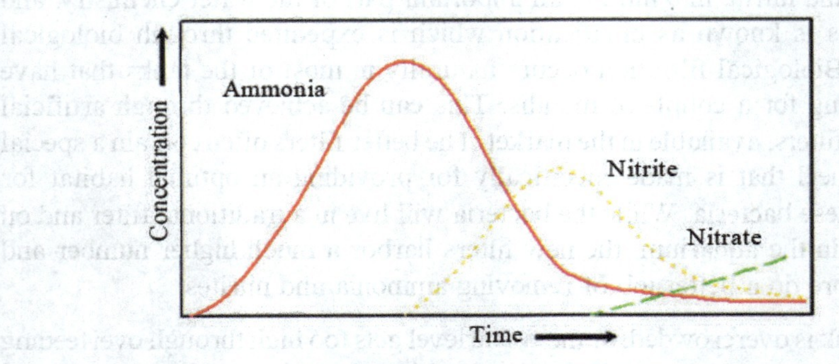

Fig. 72: Pattern of nitrogenous waste filtration over time

Water Temperature

The fish are cold blooded aquatic animals. Cold-blooded animals take on the temperature of their surroundings. They are hot when their environment is hot and cold when their environment is cold. In hot environments, cold-blooded animals can have blood that is much warmer than warm-blooded animals. Cold-blooded animals are much more active in warm environments and are very sluggish in cold environments. This is because their muscle activity depends on chemical reactions which run quickly when it is hot and slowly when it is cold. A cold-blooded animal can convert much more of its food into body mass compared with a warm-blooded animal. It remains same as the water temperature around them. A fish if not kept in their normal temperature range, they get stressed and die after getting infected with diseases. The majority of the fish are tropical in habitat that means they come from tropical climates with water temperatures around 75 ° F. Even cold water species such as goldfish cannot tolerate very cold water or sudden fluctuation in temperature. Therefore, it is advisable to understand the fish and their temperature tolerance limits.

Temperature is one of the most important environmental factors controlling and governing the metabolism of animals. Water has a high thermal capacity compared to the air and capable to absorb a large amount of heat energy if there is a small rise in temperature. Therefore, it provides a thermally stable environment.

In an aquarium, sudden change of temperature should be avoided. The thermal shocks are most likely to occur when the fish are transferred from one tank to another or during their entry to the aquarium complex. The simple management practice is to keep the fish packets floating for some time before transferring the fish in their new tanks until the temperature has equilibrated or alternatively to slowly mix the water in the plastic bags with the same water of the tank for half an hour or more. Increase in temperatures has the most distressing effect since respiration rate and excitability increase while the oxygen-carrying capacity of the water decreases. Such increase in temperature may result from refrigeration breakdown or thermostat malfunction in the tanks. The damage caused by the faulty thermostat in the system can be minimized by installing the minimum wattage heaters for temperature control or by including a high temperature cut-out in the circuit. This could simply be a second thermostat in series with the first, but set to a slightly higher temperature so that it is kept on continuous during normal operations. However, it is important such thermostat should be serviced regularly to ensure it does not stick on and work properly.

During winter months, it is necessary to provide heating arrangements to the tropical aquarium fish. A water heater of capacity 5-6 watts is required per gallon of water. Heating equipment of the aquarium contains an electric heating coil with thermostat fitted in a glass tube to control temperature. The glass tube containing the heating arrangements is submerged in the aquarium, connected to the electricity supply and the built-in neon indicator. A thermometer should always kept inside the aquarium to monitor the temperature. It should be placed at the front so that can be easily seen. The heater and thermostat should be always placed at opposite corners of the tank to get correct temperature readings of the aquarium water.

Water hardness

The definition of water hardness is often confusing and therefore, many aquarium owners overlook its importance. Water hardness is important because it is closely related to pH. As the fluctuation in pH should be minimized to protect the fish from stress similarly levels of water hardness that they thrive in, is also important to be maintained to avoid the stress. If the hardness is not optimum, it can cause not only stress but mortality also. Water hardness is the level of minerals available in the water. Hard water has considerably high dissolved minerals, and soft water has very little dissolved mineral. The most common mineral in water is calcium. Mostly tap water is either slightly hard or soft depending on the source from where it comes. The water from the well constructed in the areas that have a lot of limestone (calcium) is often hard. Water that comes from lakes, where rain water is harvested is often devoid of mineral and is soft. It is important that one should know the hardness of the water that they use in their fish tank. There are some ornamental fish species which prefer hard water and some love soft water. It should be in the range of 100 to 250 mg/L.

There is a second reason that makes hardness important as it affects pH, hard water that contains high mineral content is usually high in pH and soft water with low mineral is usually low in pH. The mineral in hard water acts as a buffer which reduces the amount of acid in the water. The resulting water will be more alkaline and higher in pH. The problem arises when it is tried to lower the pH in hard water. If a commercial pH reducer is added to an aquarium that is filled with hard water the mineral in the hard water will buffer the water and make it difficult to lower the pH successfully. The mineral from the water should be removed before effectively lowering the pH. The same is true for trying to raise the pH in acidic water that is soft and does not contain much mineral. Until mineral are not added to the water, it will be difficult to successfully alter and maintain a high pH level. So, mineral should be added in the form of calcium-based rock, to make soft water hard and more alkaline (higher in pH).

To soften hard water, it is required to remove the mineral from water using the water softener, reverse osmosis or a specialized chemical that irreversibly binds up the mineral. Another option is to find a source of demineralized water for the fish tank. Rain water is soft in nature and best for aquarium fish that prefers soft water.

Of course, there are some more alternative for this purpose and can be selected based on the fish and plant species of tank and existing water source. For beginning aquarist this may be the best solution. There are a wide variety of tropical fish available and it is not difficult to find at least a dozen different species for every different types of water. Any decent book on aquariums and tropical fish will list the individual pH and hardness requirements of the different fish species. If the water is too hard for specific application (*such as breeding of certain species*), simply it should be mixed with de ionized water until the required hardness is obtained. Most hobbyists don't feel the need to measure this particular water quality but awareness should be created to test the water hardness before stocking the fish.

Chlorine

Chlorine and chloramine are found in municipal water and must be removed before a fish is introduced in an aquarium. Chlorine is commonly added to the tank water to disinfect it by killing bacteria and pathogens. Excess of chlorine is harmful to fish. It may be lethal to fish at 0.2-0.3 mg/L. A dechlorinator to clean any water before adding it to the aquarium should be used. It can be reduced by adding chlorine eliminators such as hypo @ 1 g per 5 lit of water or by aerating the water tank over the night.

Phosphates

Tap water also contains a salt called phosphate, which serves as a nutrient to grow algae. The primary input of phosphorus into aquaria is from fish food and unfiltered tap water, whereas the primary output is through water changes, uptake by macro algae and the use of absorptive chemical filter media. Though it is essential for all forms of life, phosphate is required in rather small quantities. This poses a distinct problem in aquarium systems, specifically because the rate of input can greatly exceed the sequestration/export rate. Maximum level of phosphate tolerance in fish is 2-3 mg/L. But, less than 0.05 mg/L is ideal. Reef tanks should be maintained at less than 0.05 mg/L phosphate.

Water quality management in aquarium

The most important parameters in aquarium tank/farm management is maintaining the water quality preciously. There are many excellent products available in the market to assist water quality management, but still there is no replacement for water exchanges, good filtration, and good feeding habits.

Water exchange

Frequent change of water in small quantity is preferable to larger but less frequent changes. The water quality parameters must be in optimum range with tolerable pH and temperature. Weekly water changes are probably the most important part to maintain good water quality. Weekly water changes of around 15-20% of the total water volume improves many existing problems in water quality. The water exchanges bring fresh mineral rich water into the tank. The fish, plants and bacteria utilize the trace minerals in the water and by adding new water frequently, minerals also get replaced. By replacing water the excess of nitrate and ammonia gets removed that builds up in the water as well. It also helps to remove other toxins or pollutants that originate in the tank. If siphon with gravel cleaner is used, the gravel must be cleaned and left over food, died fish and plant waste should also be removed. This keeps the ammonia levels down and the water clean. One should be remember that most tropical fish live in environments where currents or rainfall regularly bring fresh water and remove waste. Weekly water exchange simulates natural and optimum requirement. An important note about water exchange is to make sure that the total water exchanged does not exceed one third of the water volume. It is also important that the water that is added in tank is free from chlorine. Otherwise a declorinator (sodium thio sulfate) can be used, if chlorine or chloramines are present. Wonder shells remove chlorine, stabilize PH, and add electrolytes. Water exchanges are important for nitrate removal and build up of toxic organic and inorganic materials.

Routine partial exchange of the water is necessary to dilute the build-up of soluble materials (due to accumulation of fecal and unfed materials). Evaporation losses should be replaced by suitable water (artificially prepared or natural). The cleaning can be done by hand also but hands should be washed with shop, after the cleaning but not before because of the danger of introducing soap into the water. Scraper should be used for washing off algae on glasses. The toys, air stones and other equipments, which have the algae and other sediment attachment, should also be cleaned. Plants should be trimmed and decaying leaves should be removed regularly at the time of water exchange.

Sound and Vibration

It is often forgotten that many fishes are acutely sensitive to sound and mechanical disturbances of the water. Though the hearing of most species is restricted to low frequencies (below 3 kHz for nearly all fish, and below 1 kHz for most), at these low frequency and amplitude many species produce sounds especially during courtship.

The aquarium is often a very noisy place, with underwater noise levels in aquarium tanks often very higher than those in the sea or in freshwater. Much of the noise comes from the machinery; pumps and compressors associated with the aquarium, and characteristically contain strong signal frequencies in its spectrum. Human footfalls, doors opening and closing etc. can also be troublesome, and their strongly impulsive nature may evoke startling response from the aquarium inhabitants. Vibration is transmitted to the water, mainly through the floor and tank supports but also through the water pipes. Therefore, machinery; pumps and compressors, which are producing unnatural sound, must be replaced at the earliest.

Measures to maintain good water quality

Aquarium should be filled with clean portable water. The requirement of quality water in the aquarium depends on the types of the fishes being kept there. The tap water is probably the safest source for majority of tropical fish and plants. But it contains chlorine, which is toxic to fish even at low concentration. To remove the chlorine naturally, it is better to allow maturing of the water for few days or aerating it overnight before addition. During emergency conditions dechlorination can be done with the commercially available chemical (sodium thiosulphate), purchased from pet shop.

The degrees of hardness have several biological effects upon aquatic life. Bicarbonates tend to prevent a solution from changing in acidity. Soft water, lacking this protection, may become particularly acidic if much carbon dioxide is present; such a change creates stress for organisms. For soft water species excessive hardness causes an organism problem in absorbing substances through its delicate membranes. This is mostly happen to the sensitive naked cells of eggs and milt, so that soft water has been found to play a vital role in the successful reproduction of many species of freshwater fishes. Thus, at least for the purposes such as fish breeding, a soft solution is desirable.

To maintain soft water, all sources of calcium carbonate such as calcareous rocks, gravels, coral, broken shell and algae must be kept out of the aquarium system whilst using only soft water initially and during exchange. Conversely, presence of such sources will preserve the water hardness. Some of the important water quality parameters and their optimum ranges for maintaining the fish in aquarium are presented in the table below (Table-4).

Table 4. Water quality parameters for an aquarium

Parameter	Range
Temperature	17-38^{0}C
pH	7.0-8.5
CO_2	< 5 ppm
Alkalinity	75-250 ppm as Ca CO_3
Hardness	50-200 ppm as Ca CO_3
Dissolved oxygen	5.0-10.0 ppm
Free ammonia	<0.05 ppm
Ionised ammonia	<0.1-0.4 ppm

Table 5. Ideal water quality parameters for fish breeding

S.No.	Name of the fish	Water Temperature (°C)	pH	Water hardness (mg/L $CaCO_3$)
	Egg layers			
1.	Gold fish (winter breeder)	18 - 20	7-7.5	90-200
2.	Koi carp (winter breeder)	20 - 22	7- 7.5	70 - 200
3.	Angel (summer breeder)	22 – 32 (breeding) 28 – 30 (larval rearing)	6.3 – 8.5	70 - 200
4.	Gourami (summer breeder)	24 - 30	6.0 – 7.0	60 - 100
	Live bearers			
5.	Sowrd tail (summer breeder)	28-30	6.5 – 7.5	80 - 250
6.	Platy (summer breeder)	28-30	6.5 – 7.5	80 - 250
7.	Guppy (summer breeder)	28-30	6.5 – 7.5	80 - 250
8.	Molly (summer breeder)	28-30	6.5 – 7.5	80 - 250

Precautions

Adjustment and/or corrections to existing water chemistry must be made gradually. Stability is as important as the quality of water.

Softening of Water

Some fish (e.g., discuss, cardinal tetras, etc.) prefer soft water. Although they can survive in harder water, they are unlikely to breed in it. Thus, it is essential to soften the water despite the hassle involved in doing so. Water hardness is of interest to aquarists for two reasons: to provide the proper environment for the fish and to help stabilize the pH in the aquarium. There are two types of water hardness: general hardness (GH) and carbonate hardness (KH). A third term commonly used is total hardness which is a combination of GH and KH. General hardness is more important of the two in biological processes. When a fish or plant is said to prefer "hard" or "soft" water, this is referring to GH. Incorrect GH will affect the transfer of nutrients and waste products through cell membranes and can affect egg fertility, proper functioning of internal organs such as kidneys and growth. Within reason, most fish and plants can successfully adapt to local GH conditions, although breeding may be impaired.

Typical home water softeners or demineralizers soften water, using a technique known as "ion exchange". They remove calcium and magnesium ions by replacing them with sodium ions. Although this does technically make water softer, most fish won't notice the difference. That is, fish that prefer soft water don't like sodium either, and for them such water softeners don't help at all. Thus, home water softeners are not an appropriate way to soften water for aquarium use.

Pet shops sell "water softening pillows". They use the same principal of ion-exchange. Once "recharge" the pillow by soaking it in a salt water solution, and then it can be placed in the tank where the sodium ions are released into the water and replaced by calcium and magnesium ions. After a few hours or days, the pillows (along with the calcium and magnesium) are being removed, and the pillow get recharged. The pillows sold in stores are too small to work well in practice, and shouldn't be used for the same reason cited above.

Peat moss softens water and reduces its hardness. The most effective way to soften water via peat is to aerate water for 1-2 weeks in a bucket, containing peat moss. For example, in a (plastic) bucket of the appropriate size, adding a large quantity of peat (a gallon or more), should be added one boiled (so that it sinks). It needs to, stuff in a pillow case, and place it in the water bucket. An air pump should be used to aerate it. In 1-2 weeks, the water get softer and more acidic. This aged water can be used when making partial water changes of the tank.

Peat can be bought from pet shops, but it is expensive. It is much more cost-effective to buy it in bulk at a local gardening shop. Although some folks place peat in the filters of their tanks, the technique has a number of drawbacks. First, peat clogs easily, so adding peat isn't always effective. Second, peat can be messy and may cloud the water in the tank. Third, the exact quantity of peat is required to effectively soften the water is difficult to estimate. Using the wrong amount, it results in the wrong water chemistry. Finally, when water is changes, the tank's chemistry changes when new water is added (it has the wrong properties). Over the next few days, the chemistry changes as the peat takes effect. Using aged water helps to ensure that the chemistry of tank doesn't fluctuate while doing water changes.

Hard water can also be softened by diluting it with rain water, distilled water or R/O water. Rain water is a good and cheap source of soft water. Rain water should be harvested in fibre or cemented tanks. However, if there is any factory or industry nearby to aquarim units and it is continuously emiting smoke in the sky and polluting the environment, then one should be careful. In such cases, rain may be very acidic and corrosive. R/O (reverse-osmosis) is purify water through an R/O unit. Home drinking water filtration units having R/O system costing (around Rs. 10000/-) can also be purchased from stores. Unfortunately, R/O units are slow and running a large unit with R/O system may be expensive for most hobbyists. The same applies to distill water system as well.

Aquarium filters

Power Filter: It is very easy to maintain power filter hanged on the back of the aquarium (easy access). Water is sucked through a process of mechanical filtration, using floss and insert cartridges. They also provide enough space for chemical filtration media. Within the last few years a wet dry wheel (bio wheel) is developed, to provide an even larger area for bacteria to settle. It should be wash once in a week. It Costs: Rs. 400-1000/-

Corner Filter: Water is forced through it. Inside filter floss or other media are seen. It is mainly a physical/mechanical filter. Beneficial bacteria settles on the medium and provides biological filtration. This very inexpensive filter is an excellent way to set up a hospital tank. It works as biological and mechanical filter for hospital tank.

Canister Filter: It is basically an enhanced corner filter. It is a closed box where water is forced through filtration media (mechanical and/or chemical). It can be placed inside the aquarium, or outside (underneath the aquarium or as hang on type). The canister filter has the most powerful mechanical filtration system, and can be used with messy eaters. The flip side is that it requires frequent cleaning. Bacteria also settles in this type of filter. Biological filtration can be improved, by placing wet dry wheels at the outflow of the canister filter.

Under Gravel Filter: The undergravel filter (UGF) is basically a perforated filter placed below the gravel. Water is pumped upward through the gravel by air bubbles, water stream, or a combination of both. This slows down flow of water and oxygen allows bacteria to colonize in the gravel. The UGF is an aid for biological filtration. It does not remove larger waste particles. It has to be well maintained, especially through vacuum of the gravel. UGFs are inexpensive, but have a tendency to clog up. It is recommended to replace this filter as they age. Of course, they can be combined with a power head as a pre-filter for larger particles.

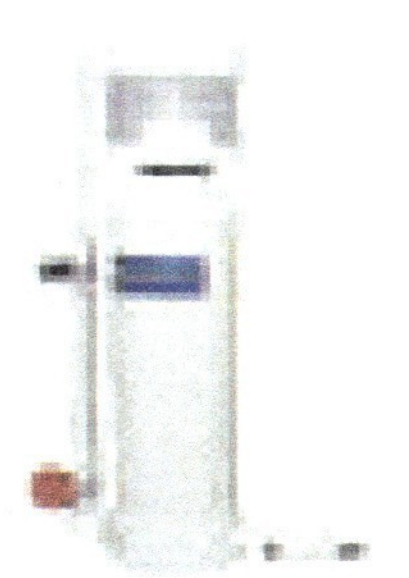

Protein Skimmer: The protein skimmer is a chemical filtration method. It takes out dissolved biological waste before it decomposes. This is achieved by a tubular design with air bubbles inside. The waste is attracted to the surface of air bubbles, which rises to the water surface. In this filter a skimmer removes the biological waste. This filtration type has revolutionized reef tanks. It only works with high pH and salinity. This filter is for salt-water use only.

Sponge Filter: A sponge filter looks like a tube with a sponge like material inside it. As water flows through, bacteria colonize in the porous foam and establish a biological filtration. These sponges also serve as a mechanical filter, removing larger particles from the water. The advanced versions of the filter use two sponges, making it easier to preserve bacterial colonies by replacing the sponges at different times. Using a sponge from an established aquarium can also jump-start a new tank or quarantine/hospital tank.

Fluidized Bed Filter: This filter is a recent development, using sand as a bacteria settlement media. In a tubular designed filter, sand is fully submerged in water. The water is pumped upwards through the sand, allowing bacteria to settle within. Additional tubes can be used as pre-filters (mechanical) and also for chemical filters using activated carbon. This filter provides a large surface for bacteria colonies, but sometimes lacks in providing enough oxygen for their performance.

Wet–Dry Filter: This filter provides no mechanical filtration but works on the principle of the wet dry wheels. It is also known as trickle filter. This kind of filter was designed with the consideration of oxygen demand of beneficial bacteria. It consists of a plastic tube with a submerged media (floss, bio balls etc.) over which water trickles, hence, it is named as "Trickle Filter". The wet dry filter provides large air water surface. The larger the surface of the media the better it works.

7

Breeding and Seed Production of Exotic Ornamental Fish

One of the attractions of ornamental fish keeping is to multiply the fishes in captivity. Although techniques used for ornamental fish breeding are not much difficult, the breeding methods for specific ornamental fish species are closely guarded secrets. Farmers have operated almost entirely on their own, developing their own methods and rely on many years of experimentation. In India, the most important breeding centres for ornamental species are located in Kolkata, Mumbai, Chennai and Kochi. Ornamental fishes are broadly classified into two groups based on their breeding methods- Egg layer or oviparous and live bearer or viviparous species. Oviparous fishes lay floating, adhesive or non-adhesive eggs that may be scattered, laid in bubble nests or deposited on substrates or in shallow pits. The breeding and rearing of live bearers are easier than egg layers, which begin with handling the live bearers.

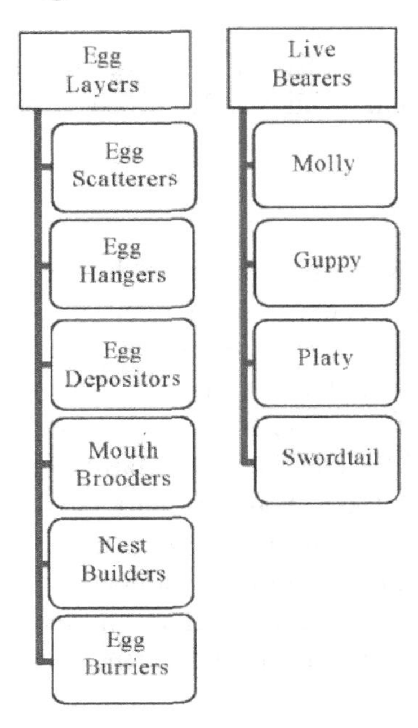

Sexing a fish

In order to breed a species, the aquarist primarily needs to be able to distinguish between the sexes. Determining the sex of a fish is an important step in knowing whether one has a pair. The sex can be easily distinguished by primary (shape of sex organs) and secondary differences (size, shape, color [sexual dichromatic], fin development). Males are frequently more colorful, larger, and have more elaborate fins. In some species, the males are slightly larger and the females are slightly more round in the belly.

Selection of the parent fish

Once males and females have been distinguished, a suitable pair or spawning group should be chosen. There are several important traits to be noted while choosing the parent fish.

- Choose the fish that display good markings like strong coloration, good fin development, etc., that should produce attractive young ones.
- Use only healthy fish for spawning because unhealthy fish, if they spawn, may produce unhealthy or deformed young ones.
- Be sure that the pair is compatible. Many species cannot be put together in a breeding tank and expected to get young ones. In fact with many cichlids, pairs form only after a group has been raised together for months or years. In certain species, one partner will bully the other to death if there is no compatibility.
- Make sure that the pair consists of the same species because hybrids are usually sterile. With some cichlids and Killifish, females of different species look similar.

Conditioning the fish to breed

In the wild, breeding is stimulated by a change in the environmental surroundings of the fish, and this can be created to some extent in aquaria. Presumably, the circumstances that trigger breeding are multifactor, consisting perhaps of a combination of factors such as food availability, water temperature, length of daylight, and changes in water chemistry. A varied diet, with an increased level of protein is recommended for conditioning. Many species can be conditioned using a well-balanced flake food, though conditioning with live foods such as brine shrimp, insect larvae, and flying insects gives better results. A small increase in the ambient temperature can prove to be beneficial, while more lighting proves a stimulus for coldwater fish species, which are normally exposed to seasonal changes. The condition of the water is significant and introducing a pair to a fresh tank may produce success. If possible, the male and female should be separated three weeks before and re-introduced at the time of breeding. For the fishes, which prove reluctant to breed, it has been possible experimentally to inject them with specific hormones to stimulate reproductive activity, but such techniques are not available to the average aquarist.

Breeding the fish

Different groups of fishes reproduce in different ways. An understanding of how the various species go about breeding is indispensable to undertake breeding programme. In general, the fishes can be divided into two broad categories – Egg layers and Livebearers. Within this basic grouping, different species have their own ways of ensuring the survival of at least a proportion of their offsprings.

Live bearers

- The members of the family *Poeciliidae* are freshwater live bearing fish distributed in the Southern USA, Central America, West Indies and northern Argentina.
- The fish are distinguished by marked differences between sexes.
- The males are usually smaller than the females and have a developed external mating organ known as gonopodium (transformed anal fin) with the aid of which the male projects sperm (spermatophores) into the female's oviduct.
- Sperms are preserved in the oviduct for a long time so that female may give birth several times in succession without the presence of a male.
- Fertilization of the eggs is followed by the development of the embryos in the ovarian follicles where they obtain nourishment from the yolk.
- There is no direct connection between the embryos and the body of the female and no nourishing substance passes from the mother to the developing embryo. This is termed as Ovoviviparity (seemingly live bearing).
- The embryos are enclosed within a thin egg shell, which burst as soon as they are expelled from the mother's body. The newly born fry are relatively large and greatly developed. Many species are variable in coloration and a great number of species interbreed.

The live bearing fishes are the easiest one among all aquarium fishes to breed; indeed, the only problem usually encountered is that of saving the young ones from the cannibalism of their parents. Various traps have been designed for the relatively rapid separation of the young ones from their mother at birth. The most satisfactory arrangement is a screen of mosquito netting on a stainless steel or wooden frame, which can be wedged across the tank so as to confine the female to one end while allowing the young ones to pass. Despite all these devices, the more natural method is having plants in abundance to provide shelter for the young ones and removing the mother at the earliest. The best plants for young livebearers are masses of *Myriophyllum, Ambulia, Nitella, Utricularia, etc.*

Fig. 73: Live bearer ornamental fishes

Fig. 74: Breeding of live bearer ornamental fishes

- Livebearers are the easiest of all the aquarium fishes to breed.
- Fertilization is internal.
- Many are cannibalistic in nature.
- Brooders are kept in the wire meshed breeding traps, through which young ones escape when born.

- Planting large number of bushy plants provides safety to the young ones.
- Presence of gonopodium differentiated male and female.
- Females, ready to breed, can be identified with their bulging belly.
- Need slightly acidic water for breeding
- Thrive well in large well planted tank.

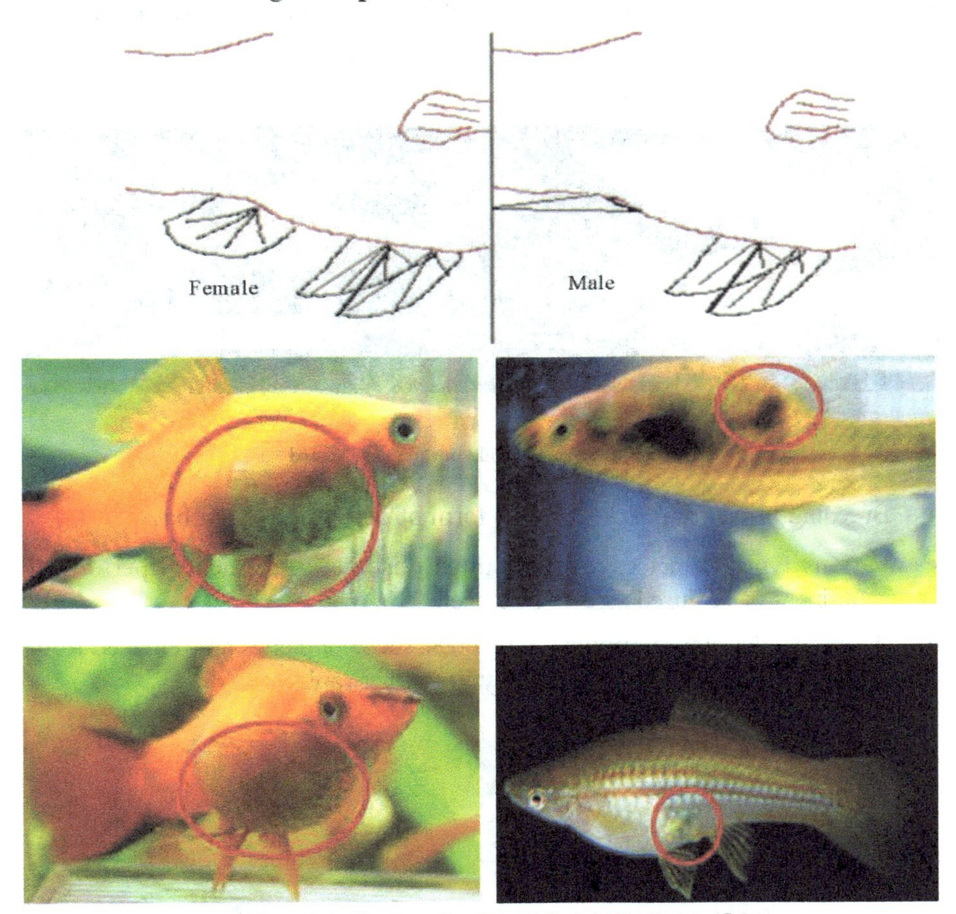

Fig. 75: Identification of male and female live bearer fish

- Temperature – 20 to 25°C.
- Gravid females are removed from community tanks as soon as they start showing swelling/belly and placed in breeding tanks either as pairs or individually.
- Breeding tank is provided with plants like Cabomba or Hydrilla.
- A tank of 100 cm×100 cm×60 cm size is ideal for mass breeding.

- Maturation -Adult platies and swordtails take 6-8 weeks and 12-16 weeks for mollies.
- Gestation period usually four weeks.
- Hardy species, breed well in most type of water but not in too soft or too acidic water.
- Many mollies appear to benefit by addition of little marine salt or common salt to water (0.5 g/l).

Fig. 76: Male and female brooders of sword tail

- Breeding traps are provided in the form of perforated nylon bags.
- Various types of net cages, perforated dustbins or fabricated perforated containers of required size can be used.
- When females stop dropping young ones; the young ones are removed from breeding trap and reared separately in rearing tanks.

Fig. 77: Perforated cages for breeding of live bearer ornamental fish

Fig. 78: A pot for breeding of live bearer ornamental fish

Livebearer young ones are quite large, and can feed on dry or other prepared food straight way. If they are given only prepared food, growth will be poor, but a mixture of live and dry food is quite satisfactory. In the early stage, feeding of live food is very important for good development. Later, it does not matter much, although the fishes will still do better with a good proportion of live food. Suitable live foods are micro worms, newly hatched brine shrimp, shredded earthworms, daphnia, newly hatched mosquito wrigglers or shredded white worms. Suitable dry foods include any fine powder food, such as dried shrimp finely ground, fine cereals, and liver or egg powder.

Egg layers

Most aquarium species are egg layers with external fertilization. Within this group, fishes can be divided into five sub-groups - egg-scatters, egg-depositors, egg-buriers, mouth-brooders, and nest-builders; depending on how they lay and handle their eggs. The preparation of spawning tank needs special care for different groups of the fishes.

i) Egg-scatters

These species simply scatter their adhesive or non-adhesive eggs to fall to the substrate, into plants, or float to the surface. The egg-scatters either spawn in pairs or in groups. There is no parental care and even they eat their own eggs, so large amounts of eggs are produced. The Characins and Cyprinids lay their eggs this way.

The spawning tank has to be set-up so that the eggs fall out of the reach of parents, because egg scatters often eat their own eggs. For egg scatters like Barbs and Danios, which lay non-adhesive eggs, the spawning tank can be furnished with a substrate consisting of two layers of marbles or a nylon netting just above the tank floor. As the eggs are laid, they fall through the marbles or the netting out of the reach of the parents. After spawning is over, the eggs or the parents can be removed.

For egg scatters that lay adhesive eggs like Tetras, the spawning tank should be furnished with a substrate. The tank should be planted with fine-leafed plants. The eggs are laid amongst plants, and adhere to the fine-leaves. The parents should be removed after spawning.

ii) Egg-depositors

In this case, the eggs are either laid on a substrate, like a stone or plant leaf or even individually placed among fine leafed plants like Java moss. Egg- depositors usually lay less egg than egg-scatterers. Egg-depositors fall into two groups: those that care for their eggs, and those that don't care. Among egg depositors that care for their eggs are cichlids and some catfish. Cyprinids, various catfish, and Killifish make up the majority of egg-depositors that do not care for their young ones. These species lays their eggs against a surface, where the eggs are abandoned. These species do not usually eat their eggs.

For those egg-depositors that care for their young ones, the parents can remain in the tank after spawning. Substrate spawners, depending on the species, should be given a tank furnished glass panes, broad-leafed plants, or flat stones for spawning sites. Some species such as Discus and Angelfish prefer vertical surfaces. For cavity spawners, flowerpots turned on their side, coconut shells, and rocky caves are suitable spawning sites. The tank should be furnished with either live or plastic plants to give the fish a sense of security.

Egg-depositors that do not care for their young ones should be given a tank furnished with fine and broad-leafed plants, Java Moss, or artificial spawning mops. After spawning, the parents or plants with the eggs should be removed. If the plants containing eggs are removed, new plants should be placed in the tank for future spawning.

iii) Egg-buriers

Fishes in this group usually inhabit waters that dry up at some time of the year. The majority of egg buriers are annual Killifish, which lay their eggs in mud. The parents mature very quickly and lay their eggs before dying when the water dries up. The eggs remain in a dormant stage until rains stimulate hatching.

A peat-moss substrate is one of the best substrates for egg-burying species. The peat moss can be removed after spawning and placed in a plastic bag to be stored for weeks to months (depending on the species). A new peat-moss substrate can be placed in the tank for further spawning. In order to initiate hatching, the stored peat can be immersed in soft water.

iv) Mouth-brooders

Mouth-brooders carry their eggs or larvae in their mouth. Mouth brooders can be broken up into ovophiles and larvophiles. Ovophiles or egg-loving mouth-brooders lay their eggs in a pit, which are sucked up into the mouth of the female. The small number of large eggs hatch in the mother's mouth, and the fry remain there for a period of time. Many cichlids and some labyrinth fish are ovophile mouth-brooders. Larvophile or larvae-loving mouth-brooders lay their eggs on a substrate and guard them until the eggs hatch. After hatching, the female picks up the fry and keeps them in her mouth. When the fry can feed for themselves, they are released.

Ovophile mouth-brooders can be bred in the main aquarium because the eggs are protected in the mouth cavity. However, it is better to separate mouth-brooders with eggs because of their potentially aggressive behavior. There are no special breeding tank requirements other than the usual tank set-up for the species.

Larvophile mouth-brooders should be placed in a separate breeding tank because the eggs are not protected in the mouth, but laid on a surface where they are open to predators.

v) Nest-builders

Many fish species build some sort of nest for their eggs. The nest ranges from a simple pit dug into the gravel or the elaborate bubble nest formed with saliva-coated bubbles. The Gouramis, Anabantids and some catfish are the most common of this type of spawners. Nest builders practice brood care.

Nest-builders should be provided with material with which to build their nests. For bubble-nest builders, fine leafed and floating plants should be provided, and the tank should have no water current to disturb the nest. Species that build nests in the substrate should be given fine gravel or sand.

Raising the fry of egg layers

When the eggs hatch, the larvae that emerge look nothing like the parent fish. Instead, the larvae have a large, yellow yolk sac and are barely able to swim. The larva feeds on the yolk sac until all the yolk is absorbed. Once the yolk sac is absorbed, the fry starts feeding on external food. The fry of small fish can be first fed with infusoria, "green water," or egg yolk. Later these fry can be fed upon larger foods like white worms, Daphnia, *Artemia* nauplii, and ground flakes. These foods are good as a first food for slightly larger fry such as those of cichlids. Once the fish grow larger, larger foods like brine shrimp, larger Daphnia, flakes, insect larvae, and chopped *Tubifex* worms are accepted. The fry should be fed several times a day. Many species need periodic sorting by size, so that larger fish do not cannibalize smaller fish.

I. Breeding of Gold Fish (*Carassius auratus*)

Common gold fish, *Carassius auratus*, belongs to the order: Cypriniformes and family Cyprinidae. It is an omnivorous fish and feeds on a wide variety of live food and accepts artificial feeds also. Colour of gold fish ranges from red, orange, silver, black, brown to white and many more. More than 30 varieties of gold fish are available. The most common varieties are Oranda, Lion head, Fan tail, Telescope eye, Bubbles eye, Albino, Pearl scale and several others.

Red cap

Fantail gold fish

Bubble eye gold fish

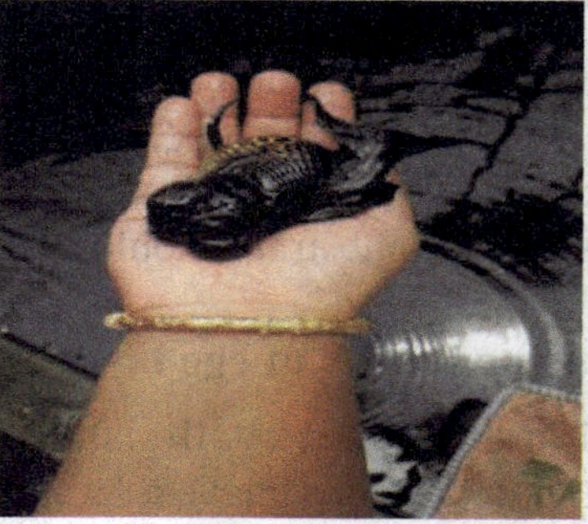

Black moore

Ryukin gold fish

Subunkin gold fish

Fig. 79: Different varieties of gold fish

Required water parameters are: pH (7.0 to 8.0), temp (25 to 30°C), dissolved oxygen (5 to 7 ppm), dissolved free carbon dioxide (0 to 4 ppm) and total alkalinity (80 to 100 ppm). Electric aerator (pump) raises dissolved oxygen level of water to 6-7 ppm which is necessary for breeding. Partial water exchange (25 to 30%) is very much essential from breeding tank. Breeding carried out in "Gamlas", 4 to 6 l capacity which is made up of clay or cement or of rectangular glass aquarium of 50 l capacity. General fecundity of gold fish ranges from 500 – 700 depending upon the size. Sex ratio is kept 2:1 (male:female) to ensure successful breeding. Eggs are generally released during night hours. Fertilized eggs are transparent and greyish in colour and unfertilized eggs are transparent white. Eggs are sticky in nature; substratum may be maintained with soft weeds, tiles, corals etc., for settlement of eggs.

Fertilized eggs hatch in 4 to 7 days depending upon water temperature. No parental care is seen. Parents eat hatchlings. As a result parents are removed after breeding.

Sex determination

In case of male gold fish white bumps or tubercles develop on the operculum and pectoral fin. Main ray of pectoral fin have thick edge in case of male but thinner edge in case of female. Fins become more pointed in case of male but look rounded in female. Vent assumes a concave shape with a small opening in male and vent becomes convex and large opening in case of female. Abdomen is seen to be smaller in male but large in case of female.

Culture of common gold fish

The culture of common gold fish is being taken up normally in cement tanks of dimension 10' x 5' x 2' or 12' x 6' x 2'. Preferable temperature for culture is

15.5 to 24°C. pH range 7.0 - 8.0 and moderate hardness of 50 – 75 mg $CaCO_3$ per litre and oxygen level of 5 to 7 ppm. Generally, 300 fry (23 mm) of gold fish are stocked in each cement tank of dimension 10' X 5' X 2'.

The newly hatched young ones depend upon their yolk sac as a food source for a couple of days. When the fry become free swimming, they are fed with *Artemia*, *Daphnia*, *Moina*, Tubifex worm and other planktons. Young ones of 2–3 days old feed on egg yolk and dried milk powder. After 10 days, the young ones start feeding on tubifex worms and maintained till their disposal.

Fig. 80: Gold fish

Fig. 81: Brooders of gold fish

Base for adhesive eggs

Eggs on mop (inside water)

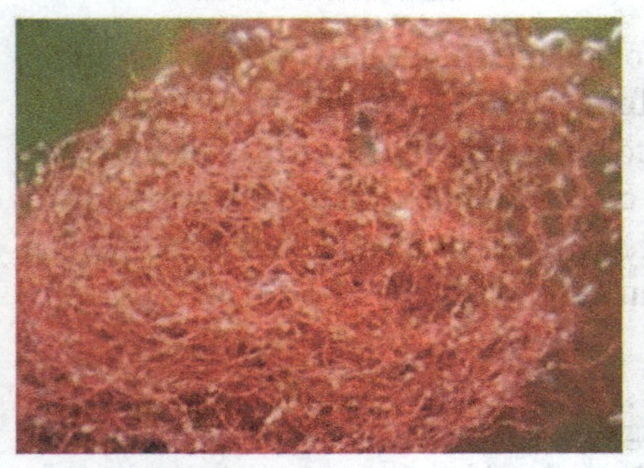

Eggs on mop (outside water)

Newly hatched young ones

Gold fish fry in rearing tank

Growth pattern

Fig. 82: Different stages in gold fish breeding

II. Breeding of Koi carp (*Cyprinius rubrofuscus*)

Koi is the Japanese word for "Love" and is also used as synonym for "fancy carps". In the Far East, koi carps are given as gifts to friends. Fancy carps, originated from China, were popularized by the Japanese. Therefore, the fancy carps of today are called Japanese carps. Among the Chinese businessmen, fancy carps are often reared for "Good Luck". In Malaysia, fancy carps were popularized in the 1970s. Its small eyes, and thick lips are the characteristics for identification. It has two barbes at each corner of the mouth that the other cyprinids lack. They have large scales and strongly serrated spines on their dorsal and anal fins. However, their color pattern is a useful distinguishing characteristic. Over their natural range, koi carps live up to 15 years, with reports of individuals living up to 24 years. Males are known to live longer than females. The koi is a hardy fish, and can be reared in small containers to large out door ponds; they quickly grow to 30 cm (1 ft) or bigger. While possible variations are infinite, brooders have been identified and named as given below:

i) Asagi - light blue on top, red or orange on bottom, blue scales bordered in white.

ii) Shusui - similar to asagi, but with large scales in a dorsal row.

iii) Bekko - primary color red orange yellow or white, with black patches.

iv) Kawari mono - miscellaneous.

v) Goshiki - mostly black, white red, white brown and blue accents.

vi) Kinginrin - bright metallic sheen, silver highlights.

vii) Kohaku - red accents on white body.

viii) Koromo - red and white overlaid with blue or silver.

ix) Ogon - uniform yellow or white.

x) Tncho - primarily white with a red patch on forehead.

xi) Platinum ogon - pure white.

xii) Shown sanke - black with red and white markings.

xiii) Utsurimono - uniformly black with red, white and yellow markings.

xiv) Taisho sanke - primarily white, with red and black markings.

xv) Tanchokohaku - pure white roxmd red head patch.

xvi) Hikarimoyo mono - two colors, one pale, one metallic.

xvii) Hikariutsuri mono - two metallic colors.

Fig. 83: Varieties of koi carp

Selection of the Parents

Choosing the right parents for breeding of koi is not difficult, but it can sometimes be tricky. Female koi carps are visibly rounder than male koi, especially those that are ready to lay eggs. Males are slimmer in appearance, and may develop roughness on their gill plates when ready to spawn. Selection of mature koi only is advisable for breeding, i.e., they should at least be two years old (younger koi will produce weak offsprings). Experts say that the optimum breeding age is 2-4 years old. Both parents should exhibit excellent body conformation and high-quality colors and markings.

Some breeders use two males for a single female during breeding to maximize the yield of the propagation. The breeding act of the koi is very physical and can harm the participants (especially the female if the males are very aggressive), so this ratio of 2:1 must not be exceeded. One advantage of using just a single male is the higher predictability of what the offspring will look like.

Preparation of the Spawning Environment

Once the prospective parents have been identified, they need to be taken out of the main pond and isolated in separate and smaller ponds where they can be conditioned for spawning. Males are separated from females to prevent indiscriminate spawning. Many hobbyists start this isolation at least 1 month before the anticipated spawning date.

Eventually the female becomes rounder and noticeably bloated with eggs. Now with a heavy but soft abdomen, she is presumed to be ready to lay eggs and is very carefully moved to the spawning pond. This egg carrying koi must always be supported by water during the move, even while inside a net. At this point the male is also assumed to be ready to participate in the reproduction as well, and is moved into the spawning pond a few hours after the female has already been acclimatized to it. Many breeders introduce the male in the evening, since spawning usually happens in the wee hours of the morning.

A Koi's prime mating age is between 3-6 years old, but koi have also been able to produce baby koi fish until they are up to 15 years old. For breeding koi, couple should be given privacy when it's time for them to mate. Koi needs a place to lay their eggs. Koi carps breed in tanks of 250-500 l capacity with weeds like *Ceratophyllum demersum* or *Hydrilla* sp., which serves as substrate for attachment of adhesive eggs, released by the fish.The pattern of reproductive cycle is dependent more on temperature than on photo period. In north India, they show two main peaks of breeding activity once during spring and again in autumn. consists of 2:1. Wild carps are partial spawners by spawning two or three times at a 14 days interval. A female koi, weighing 150 - 200 g, may lay as many as 2000-3000 eggs in a single spawning. Eggs are 0.90 -1.1 0 mm in diameter. Incubation time is 24 - 25 hours. Hatching takes place between 27 - 28.5°C. Hatchlings are fed finely sieved zooplankton. They may mature by the end of the first year.

Koi Carp Feeding

Koi are generally bottom-feeders (omnivorous and detrivorous), usually feed on artificial diet containing 70% protein (live feeds), 20% carbohydrates, 9% vegetable food and 1% minerals. The fry feed on zooplankton and when 2 cm long they prefer a diet of bottom living invertebrates and overgrown water plants. Overfeeding will lead to increase in ammonia and nitrite which if unchecked could be fatal. Certain foods can be fed to koi to enhance their color. Carotene affects the red pigmentation; Spirulina (algae) enhances the red color but won't affect the white. Live foods are taken readily which include worms, prawns and even tadpoles.

Fig. 84: Breeding of Koi Carp

III. Breeding of Tiger Barb (*Barbus tetrazona)*

Out of more than 30 commercially important species of freshwater ornamental fishes reported from Indian waters, tiger barb, *Barbuste trazona* is one of them. This species is hardier, active and does not require much of attention with regard to its basic needs. Their large scales, bright colors, schooling behavior, ease of maintenance and breeding, have made the fish popular in the aquarium trade. Though there are 1078 barb species reported from all over the world, only 70 barb species are commercially important because of their colour pattern.

Maturity of Tiger Barb

The tiger barb, which is four banded, usually attains sexual maturity at a total body length of 20-30 millimeters (2-3 cm) (or) at approximately 6-7 weeks of age. Although tiger barbs are not sexually dimorphic, males display a bright red coloration on fin rays and snout while females tend to be more round in the abdominal region and slightly less colorful. Females are usually larger than males. They can obtain a maximum length of 7 cm and body depth at 2 cm. All related barbs mate in a ratio of one male to one female with the male displaying aggressive behavior while the female is submissive.

Brood stock conditioning

Conditioning the male and female in separate tanks is an important step in the seed production process. Tiger barbs, for use as brood stock (2 to 3 cm body length), are first collected from a production ponds or natural water bodies and graded with size graders. Sexually mature females are identified by full round abdominal region and sexually matured males are identified by bright red colors on the fin rays. The selected brooders are then placed sex-wise in separate circular or square or rectangular conditioning tanks. Rectangular tanks are more conductive for removing and selecting brooders. A stocking density of one fish per four liters of water is recommended. The conditioning tank should be provided with sufficient aeration and water exchange at a rate of 20% per day. The separated fish are conditioned by feeding frozen blood or tubifex worms or Artemia. High quality flakes are given as feed at least twice or thrice per day for a period of two weeks. Since wild tubifex causes infection to broodstock, utmost care should be exercised to prevent this, through needed cleaning etc. During conditioning, good water quality should be maintained as the conditioning can lead to fouling of the water. Lack of proper conditioning will result in greatly reduced number of successful synchronized spawning.

Spawning of Tiger Barb

Submerged aquatic plants or roots are often chosen by the females as the substrate on which they deposit eggs. During actual spawning event, the male clasps the female with its fins during which eggs and sperms are released over the substrate. The behavior may last for several hours or until all the eggs are released. On an average, 300 eggs can be expected from each female per spawn. Tiger Barb consume the eggs greedily after spawning. Therefore, parents must be removed as soon as possible (Vogt and Wermuth, 1961; O'Cornel, 1977 and David, 1983). Spawned eggs are adhesive, negatively buoyant in freshwater and on an average 1.18±0.05 mm in diameter. The eggs will hatch in 3 days if a temperature of 25° to 27° is maintainted.

Induced Breeding of Tiger Barb

Sometimes, the tiger barbs does not breed under normal captive conditions. To induce these fishes to breed, hormonal intervention is needed. The hormones commercially available to induce them to breed are Ovaprim, Ovatide and Chorulon. The injection can be given inside the peritoneal cavity or just below the dorsal fin in intra muscular region. Since, the fishes are very small, care must be taken to avoid any injury to the individual fishes. Adequate care should be taken to calculate the required dosages. Now a days, some hormones are mixed with the supplementary feed and given orally to avoid any injury to these small animals during injection.

Rearing of Young Ones

The newly hatched fry do not swim for two days. They obtain nutrition from the yolk sac. So much so, the fry do not require supplemental feeding at this time. Three days after hatching, yolk sac is usually absorbed. Newly hatched brine shrimp *Artemia* approximately 0.02 inches in size are introduced as the first feed and it is used exclusively for next two days. The fry should be fed to satiation three or four times per day. After feeding with brine shrimp exclusively for two days, commercial feed can be introduced for the fry to feed upon. Once the fry gets adjusted to consume a commercial feed and consumes it for two days and are approximately 5 mm (0.2 inches) in length, they can be transferred to prepared outdoor nursery tank or directly stocked into a grow out pond or tank.

Fig. 85: Tiger barb **Fig. 86:** Male and Female Tiger barb

Fig. 87: Different stages of barb breeding

Fig. 88: Young ones of tiger barb

IV. Breeding of Angel Fish (*Petrophyllum scalare)*

Angel fish breeding has progressed into an art with the development of the veil finnages, superveil finnages and the many color varieties. It is remarkable that all of these forms came from the original standard silver angel fish from the wild.

Sex identification

It is difficult to identify male and female in the angel fish but at the time of spawning, genital papillae are the reliable sorce of identification. They look little nipple-like projections and are called ovipositors. The female's ovipositor is larger and blunter than the males which is slender and more pointed. These protuberances, which appear at the vent, are used for depositing eggs and fertilizing them.

Spawning tank

Large aquarium tank of 80 to 100 l capacity is ideal for spawning. Spawning tank water requires a slightly acidic pH level of 6.6 to 6.8. The fish can spawn at higher pH of 8 but fish tend to spawn more readily at the lower pH levels mentioned above. It is especially important to keep the water acidic if you are going to keep the eggs with the parents. The tank is furnished with slates or glass plates that are slanted at angle to lay eggs upon. An air stone, giving mild aeration, may be placed at the corner.

The pair select a spawning site and thoroughly clean it for two or three days before actual spawning takes place. When the cleanliness of spawning site finally meets the approval of the parent fish, the female will make a few test runs. She will pull her ventral fins of feelers close to the lower sides of her abdomen so that her entire lower line is relatively straight. Her ovipositor will then be able to make full contact with the slate; glass plate or whatever chosen for a spawning site. The male will then make a few practice run too before the actual spawning takes place.

When spawning actually takes place, the female will pass over the site and eggs are deposited which adhere to the surface. The male then moves in and scoots along over the string of eggs, just laid and fertilizes them, his fins taking the same position as the females so he can press closely to insure a higher fertilization rate.

The male and female angel fish pass over the spawning site until several hundred or more eggs have been laid, depending on the size and condition of the female prior to the spawning. The parents hover closely over the spawn and fan continuously with their pectoral fins to create a circulation of water over and around the eggs. Some fertilized eggs are turn white in a matter of hours and removed by the parents.

Fig. 89: Angel fish

Male and Famale Angel

Male

Famale

Eggs

Just after hatching

After 24 hrs of hatching

After 48 hrs of hatching

After 5 days of hatching

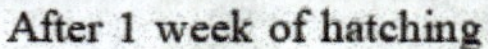

After 1 week of hatching

After 2 weeks of hatching

After 4 weeks of hatching

Fig. 90: Different stages of Angel breeding

Hatching of eggs

For the successful hatching of the eggs, it is recommended to use very soft water preferably rain water or distilled water because it has naturally low pH of 6.2-6.5.

When the spawning is over, the glass plate should be removed from the spawning tank and place it in a 30-50 l tank with sponge filter and a piece of slate leaned up against a side wall. An air stone should be placed in the jar in such a way that somewhat vigorous stream of air bubbles does not hit the eggs directly. Few drops of 10% Methylene blue is added to prevent the fungal attack on eggs.

Hatching occurs in about 36-48 hours depending on the temperature. There will be a period after hatching and before free swimming when the fry will stick together. At this time increase the aeration so all the fry will have access to sufficient oxygen.

Do not put food in the tank till the fry are free swimming. After about 3-5 days when they are free swimming, introduce newly hatched brine shrimp into the tank for the fry to eat.

In rearing tanks for baby angels that are two weeks old, incorporate normal dechlorinated tap water. Ten liter for every hundred liter of water is changed daily from the bottom of the tank where all the detritus accumulates. These rearing tanks are not treated to lower acidity.

Feeding schedule

One week to three weeks

Angel young ones do not need any type of feeding until they are in free swimming stage. It takes about four to six days depending on the temperature. When the young fry become free swimming, feed them newly hatched brine shrimp (*Artemia*) or *Moina*.

Brine shrimp is fed directly to the young ones at first to make sure that no excess food is floating around in the tank for hours at a time. Three or four feedings per day should be sufficient. Any brine shrimp floating around after 20 minutes is a sign that you are feeding too much. Remember, feeding in light quantity decreases overfeeding and associated problems such as ammonia and disease.

Three weeks to five weeks

After three weeks, the fry attain a size in which they will accept finely crushed flake foods. Flake foods are provided in small quantities as a supplement.

After three weeks, brine shrimp can be fed in such a way that they are consumed within 15 minutes of adding it to the aquarium.

Five weeks to seven weeks

After five weeks of age, the young angel fish are introduced to dry foods. A small amount is fed twice daily until the seventh week. During this time, the small angel fish will attempt to eat the dry flakes but they will usually spit it out soon after taking it into their mouths. Some will eat the flakes and some will not. Around the seventh week the angel fish begin accepting dry flakes and there should be few flakes, if any remaining on the bottom of the tank like the previous weeks.

Six weeks to adult hood

At about six weeks of age, the young angel fish reaches a size in which they will begin accepting blended beef hurt cubes. Baby brine shrimp can still be given to the young angels up to three months but beef liver and flakes are all that is necessary for quick growth.

V. Breeding of Gourami (*Trichogaster* sp.)

Gouramies although closely related to Bettas, do not have fighting nature. Under good conditions, they are friendly fish. In all gouramies, the pelvic fins are shaped as long as thread-like feelers, which can be moved in all the directions. Popular aquarium varieties of gouramies are the giant gourami (*Colisa fasciata*), dwarf gourami (*Colisa lalia*), pearl gourami (*Trichogaster leeri*), blue or three spot gourami (*Trichogaster trichopterus*), moonlight gourami (*Trichogaster microcephalus*), snakehead gourami (*Trichogaster pectolaris*), chocolate gourami (*Sphaericthys osphronemoides)* and kissing gourami (*Helostoma temmincki*).

For describing the breeding of gouramies, a typical example of blue or three-spot gourami is presented below:

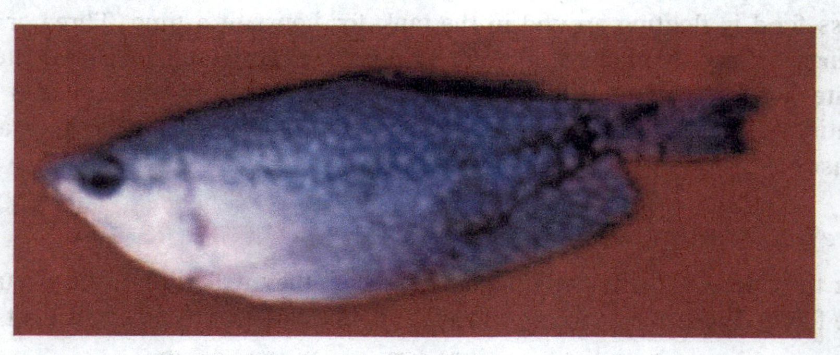

Fig. 91: Blue Gourami (*Trichogaster trichopterus*)

The three spot gourami breeds during April to August. During breeding season, mature male develops dark coloration and female shows bulging abdomen. While making breeding pair, care must be taken to select the mature female, which is ready to spawn. This is because males of blue gourami are very aggressive in nature and tend to kill female, if she is not ready to breed. Aquarium tanks of 50-80 litre capacity can be used for breeding. The water level in the aquarium should not be more than 25 cm. One or two pieces of floating plants and beetle leaves may be floated on the water surface to hold the bubble nest. The tank should not be provided with aeration. The pairing of blue gourami is made in the ratio of 1:1. If the male is in breeding condition, it will start making nest within one or two days. The bubble nests float under the plant leaves and look like soap foam. Like many anabantoids, the male blue Gourami wraps his body tightly around the female, turning her on side or back, so the eggs will rise unimpeded to the surface. This close embrace is also important because it brings the reproductive organs as close as possible. Because sperm cells survive only for a matter of minutes in the water, the timing of their release and proximity to the eggs is critical. Just before the sperm are released, the pair may be observed quivering-a sure sign that spawning is near completion. The eggs are released immediately thereafter, and are fertilized by the time they reach the bubble nest. The pair may repeat the process a number of times over the course of several hours. It is not unusual for the number of eggs produced to reach to thousands. A total of about 500-1000 eggs are produced. Once the spawning is over, the female must be removed from the tank. The male will take care of the eggs and young ones. Hatching takes place within 24 hours. As soon as the fry are free swimming, the male should also be removed from the tank.

Fig. 92: A ready bubble nest

Fig. 93: Hatchlings of gourami

After 36 hours, when the fry are at free swimming state, they are provided infusoria as feed. After 6-7 days the fry start taking brine shrimp or small *Moina*. At this stage, they should be fed 3-4 times a day. The growth of fish is very uneven and often some "shoot fry" develop. The "shoot fry" i.e. bigger ones of the lot should be separated and reared in the different tank. Now, they can be stocked in a bigger tank and give a diet of worms and formulated feed.

Breeding of fighter fish (Betta)

Bettas are vibrant colored fish with long flowing fins and are very popular among aquarists.There are hundreds of different varieties of Betta fish or fighter fish, all with different shaped tails, patterns and color. Some people choose to breed these fish to create new varieties, some choose to breed them for shows, and others want to breed a very particular variety based on tail type, color and pattern.

Fig. 94: Fighter fish

Fig. 95: Keeping males separately during breeding of Fighter fish

Spawning tank

If it is not possible to use lots of separate tanks, a divider can be used to keep the males and females separately. Whilst they are native to the shallow rice paddies in Thailand, they still have access to miles and miles of water in their natural habitat. Therefore, Betta fish tank will need to be at least of 5 gallons; need a heater and a filter, some live plants such as java moss or java fern, some hiding places and smooth gravel.

In addition to the permanent tank, there is also a need of breeding tank for Bettas. This will be a stripped-down tank compared to the permanent set up. Tank should be set up somewhere quiet; they like privacy while breeding so keep it away from busy areas and other busy aquariums. The tank should be filled with three to five inches with water. Heater and filter may be added. The breeding tank will need a heater to keep the water at around 78°F.

Some people choose to use filtration, and others don't. Using filtration will create a small current in the water which can disturb the bubble nest so if use a filter, it should not be powerful. Use of gravel is not advisable, because the eggs might settle in it, leave it bare bottomed.

It's essential to have hiding places due to the aggressive nature of the males during breeding. Plants can be used to provide hiding places; the plants also provide a place for small organisms to grow which the fry can eat. Java fern is an ideal plant to use, or plastic ones can also be used.

There will be need of something that will float on the surface of the water, to provide a surface for the male to build his bubble nest. A couple of popular choices are to use a piece of Styrofoam or an almond leaf. As for lighting, Bettas need privacy to breed and won't spawn if it's too bright, so only a dim light should be used.

It's also important to ensure the breeding tank has been properly cycled. This process takes around 4-6 weeks so always tanks should be prepared in advance.

Selection of brooders

Selection of breeding pair depends on type of betta to be bred. Fish should be selected with the characteristics that one want to breed, for example if someone wants Crown tail Betta, he will need to choose the adult fish with the colour he wants to achieve. The selected fish should be actively swimming around, with no sign of disease or infection. Betta should be free from parasites, disease and fungus. It is to ensure that they are not lethargic or lying on the floor of the tank. Their eyes should not be bulging and should be clear. Breeder should also check their scales and fins for any sign of fungus, damages or tears. The age of the fish is also important. Bettas are at their peak for breeding between four and 12 months. Older fish are still able to breed, but will have more success with younger ones. Finally, it is important to consider the size of the fish. They should be roughly of the same size, and the female should be slightly smaller than the male. Never use a large female and a smaller male.

Breeding technique

In the wild, it is the female Betta that selects the male, betta based on their size, color, fin length and the quality of their bubble nest. However, in captivity, the likelihood is that one have already chosen a pair to mate so we need to ensure the conditions are optimal. Once the tank is prepared with right conditions, the fish are conditioned so that they are able to breed effectively.

Breeding is not easy work. The fish need to be conditioned to ensure they have enough stamina and are able to withstand the stress which they will endure during courting and spawning.

If the favourable condition is provided it is more likely that the female will produce enough eggs and the male will have enough energy to care for them. One should allow a minimum of two weeks to condition the fish, more experienced and committed breeders will allow even longer.

To condition them we need to feed them high quality food in small amounts for two to four times per day. Live foods are the best to do this. Bloodworms, daphnia, finely chopped meat, tubifex worms and small insects such as crickets and roaches can be used. If there is no access to live foods, frozen substitutes will also work.

Introduction of the Female

It's time to introduce the female to the male; this is not a quick process and need to be done slowly. There are a couple of ways to add the female to the tank. Either it is introduced in one part of the same tank which is divided by using a divider or floating it in a see-through container.

She should be adjusted to new setting for around 30 minutes before introducing the male. If a divided is used, the male should be added to the opposite side of the tank and allowed are to swim around. It will be noticed that the male turns a deeper shade of color and will start to flare his fins in an attempt to impress the female. You might notice him nipping the container that she is in (this is to be expected). If the female is interested in him, her color will also darken and she will display vertical stripes across her body, known as her barring pattern. Her oviposter (a small white tube through which she lays eggs from) will also protrude behind the ventral fin. It can also be noticed that she wags her body back at him. The male will usually build a bubble nest within 24 hours of seeing the female. The bubble nest is made from air bubbles which are made by the Betta and they're coated in saliva to make them durable. The bubble nest will usually be built next to an object that breaks the water, such as the Styrofoam or the leaf. Sometimes, they are built at the surface of the water and sometimes they are built just below whatever is floating.

Release of the Female

Once the nest is made, you can release the female into the tank. This is usually around 12-24 hours after first introducing them. If a small container is chosen , female cannot live there for longer than an hour and move her back to the permanent tank while the male builds his nest. If a divider is used, the divider can be removed and the female is allowed to access the whole tank.

She will most likely swim straight to the bubble nest to inspect it. If it's not up to scratch, she will either swim away or destroy it. If she does destroy it, remove the female and start the process again. If she destroys it a second time then it is required to find another pairing. When the male has spotted that the female is within his reach, his displays will become more impressive and he will start to chase her around the tank. It is normal to see chasing and biting for a few hours, but it's still important to keep a close eye during this stage and be ready to intervene if things turn nasty.

They start to perform a dance where they swim next to each other and flare their fins every inch or so. The male will chase her and nip at her fins if she isn't responding; she may also need somewhere to hide (this is why plants or hiding places are so important).

This will continue until the female is ready to spawn. The signs which females display, indicating they are ready to spawn, depends on their nature. Some females show their submissiveness straight way, swimming up to the male with their head down and fins by their side, others will charge at the nest.

The Nuptial Embrace

Once she is ready, the mating dance will begin. The aim of the male is to flip the female upside down, and wrap himself around her to fertilize the eggs as she releases them. Once he has achieved this, they will either stay floating or sink to the bottom. He'll then release her and she'll have a few minutes to recover before trying again. After they've performed a few of these squeezes, the female will begin to drop eggs each time they embrace. This embrace is not to help her release the eggs, but to increase the likelihood that the eggs are fertilized as their ventral are closed together this way. The female may float sideways and look lifeless while she lays the eggs, but this is normal. This process can last anywhere from a few minutes to a few hours. Most Bettas lay between 30-40 eggs per spawn but some can lay up to 500. The male will start to catch all the eggs, taking them up to the nest. Some females help with this process once they're recovered, but others tend to eat the eggs so it's advisable to remove the female as soon as she has recovered.

This is also advisable because the male may attack her because he sees her as threat.

Fig. 96: Male collecting the eggs to put them in the bubble nest

Hatching Eggs

Once the female has been removed, cover the tank with plastic wrap to keep the moisture and heat in. This creates a humid environment and helps in hatching of the eggs and develop the fry. When the female has finished laying eggs, the male will release milt into the tank to externally fertilize the eggs. Over the next day or two, the male will spend his time catching any falling eggs and blowing more bubbles. Sometimes, they build new nests and move the eggs there. Sometimes, they will eat the occasional one, perhaps if they haven't been fertilized properly.

Feeding Betta Fry

Over the first 36 hours, the fry will use up all the oxygen in the bubbles, which causes them to collapse. As the fry start to hatch, they fall from the bubbles and the male will catch them and put them back. It is expected that the fry to stay in the bubbles with their tail hanging down for 2-3 days until they're ready to swim horizontally by around day 4. As soon as the fry are free swimming, the male will need to be removed. Or if you have a fry raising tank, the fry can be moved there. It can take up to four months to raise the fry to adulthood so you will need plenty of food. They will need small foods such as brine shrimp, infusoria or micro worms.

VII. Breeding of Arowana or Dragon Fish *(Scleropages formosus)*

The arowana is a very valuable commodity. These dragon fishes are particularly revered in China, where the freshwater pets are status symbols and cost a pretty penny. Arowanas are large, elongated fish with upward facing mouths. They live mainly in freshwater and have a low tolerance to saline water. The habitat has to be specific in terms of water, size of tank and more. They obtain oxygen by breathing air at the surface of the water. These fish can grow relatively long, and they have an almost eel-like elongated body. Their dorsal and anal fins are short, soft, and extremely elongated. This gives them the appearance of an extremely long tail fin. There are a number of different arowana species, each with their own unique differences. Some species come in brilliant arrays of color, from bright red to dazzling silver. The fish average life span is 10 to 15 years. These predatory fish will prey on a wide variety of food, basically anything that fits in their surprisingly large mouth. Most of their feeding occurs at or near the surface of the water. They prey on crustaceans, insects, small fish, and basically anything that falls into the water. Specimens are sometimes found with birds, snakes, and even bats in their stomach.

They breed in the wild during monsoon and are mouth brooders. Asian Arowana are hard to sex, especially, when they are young. Sexual maturity depends on environmental conditions, but generally female takes 2-3 years and male 4-5 years to reach maturity. Generally, males are larger in size and fin than its female siblings, which give them a more arrogant look. It is believed that they also have wider and deeper jaw (to hold the eggs and fry). Females, on the other hand, are smaller and have a more rounded body. It takes a lot of experience to identify sex of the mature. Matching is inherently more dangerous as Arowana are territorial. They will inevitably fight if kept together. Their aggression decreases when the number of Arowana kept together increased to 6 or more. This is often not possible due to high price of some species of Arowana. Breeding of Arowana is difficult and can be done only by professionals. In the wild, Arowana pair by natural selection spends weeks courting. A pair of Arowana will chase and bite each other's tail while courting, eventually together side by side and chasing off other fish before they finally breed. When the female is ready she will lay eggs on slow stream riverbeds, which will be up to 1/2" in diameter and are then fertilized by the male. The eggs are then scooped up by male Arowana and hatched in his mouth. The fry will begin to briefly leave the father's mouth slowly increasing their exposure to the outside world. There will be fewer Arowana than originally consumed. It is believed that the Arowana when startled sometimes accidentally swallows some of the young. During this time, the father will signal the fry when there is a sign of danger and they will immediately swim back for safety. The babies will have a yolk sack that they will use until they are ready to feed on their own. The fry will leave the father when they are capable of surviving on their own.

Fig. 97: Arowana

There are few records showing that Asian Arowana has been successfully bred in aquariums. Commercial breeders of Arowana usually use large earthen ponds. When breeding in earthen ponds, ten or more mature Arowana (half males, half females) are put into the pond and natural selection is allowed to take its course. The Arowana are observed carefully. When a pair is formed, they will chase the others away and start laying eggs. A net is then put in to segregate the pair from the other Arowana. When the fry are free swimming, they are netted out and kept in rearing tanks. However, most aquarist do not have an earthen pond at home or live in a climate suitable to leaving them outside for several months.

Natural pairing

At least six mature Arowana are needed to have a reasonable chance of getting a pair. Young, ones that have grown up together, are preferable. If not, they should be placed into ones the tank at the same time so that none will have chance to develop a territorial sense earlier than others.

Close observation is needed until a pair appears to have been formed, noted by the male swimming closely to the female and chasing the rest away. At this time removal of selected pair from the rest of the Arowana from the tank is an important activity. These fish require more intensive care than most freshwater species. They are mostly surface dwelling, so tanks should be long and wide rather than deep.

Courting will take place over a 1-2 months period. They will continue to swim closely together, often in circular motion, chasing and biting each other's tails and fins, often causing injuries to the body as well. Females usually sustain more injuries, especially in anal fin, genital area and gill cover, possibly due to male trying to stimulate her hormonal secretion.

Their appetite will drop gradually, as the abdomen of the female will start to swell up and filled with eggs. An area, usually a less disturbed area with weak current, will be chosen to lay the eggs. Covering the spawning area of the tank is recommended to avoid disturbence or termination of the spawning due to stress. At the time of egg release, they will swim parallel to each other, rubbing against each other and occasionally stay motionless. Eventually, they will stop swimming, but continue rubbing bodies until, with a sudden spasm, the female will release her eggs and simultaneously the male fertilizes the eggs by releasing the sperm. Spawning usually takes place in the afternoon with the number of eggs laid averaging 60 and being up to 1/2" in diameter. The male will immediately scoop up as many eggs as possible with the remainder usually eaten by the female. During hatching period, female start to chase the male, possibly violently. It is possible that the male might swallow some eggs accidentally during this chasing, therefore, it is a good idea to remove the female away at that time.

The yolk sac will be fully absorbed in around 60 days. The fry will then begin to leave the father's mouth. Breeder must have supply of baby guppies on hand at this stage. Once the fry begins to eat readily for a few days they can be removed from the parent.

Matching of pair manually

This is risky. One should be confident that he/she has two Arowana of the opposite sex. This takes a lot of experience and should only be tried with extreme care. Arowana can and will fight to the death if paired improperly. Conditioning of the two Arowana requires lots of live food. Two fish will be introduced into the breeding tank of at least 175 gallons with a transparent divider between them. Give them time to get accustomed to each other and continue to condition them at the same time. After several weeks, divider should be removed and observed carefully. They may be more violent than a pair formed by natural selection. When it gets too violent or fish are injured, the divider should be placed back. It will check the right selection of pair. If they are of not opposite sex or are not mature enough yet., a pair give it a few more days and remove the divider again. If the pair is compatible, the two Arowanas will show courting behavior as stated above.

In Asian culture, the Arowana is believed to closely resemble the Chinese Dragon. The fish are even believed to be reincarnations of dragons, and are considered a symbol of luck, strength, prosperity, and wealth. Now the prized fish are being bred to produce unique color combinations. Chili red Arowanas go for around $1,400; emerald-violet fusion super reds run about $12,000; and the rare albinos, while less striking, command six figures.

Fig. 98: Asian red-blue arowana

VIII. Breeding of Oscar

The oscar (*Astronotus ocellatus*) belongs to the cichlid family known under a variety of common names, including tiger oscar, velvet cichlid, and marble cichlid. Oscar Cichlids come from the Amazon River, primarily sluggish moving areas that are extremely warm. Oscars do not tolerate cold water at all, and temperature is one of the main restrictions of their habitat. They are considered to be the most intelligent aquarium fish available to hobbyists. Oscars have an impeccable memory and are one of the few aquarium pets that can be trained to do tricks and individually distinguish its talent. Oscars are often available in a few different colors. The most popular being black with bright red scale colorations through its center. Other less common variations include a bright albino white with translucent red eyes, shades of pale blue and even banana yellow. Oscars enjoy a large variety of foods. They are hailed for eating almost anything put into the tank. For this reason, it is important to feed them a balanced diet. They are often fed with feed with high in fats and can also run the risk of infecting it with disease or parasites. It is to ensure that any live fish fed to the Oscars have been properly quarantined. Live crickets and bloodworms are an excellent alternative and are more aligned to an Oscars natural diet. Live snacks should be fed in combination with commercial granules, frozen worms and brine shrimp and beef liver. This will ensure a healthy and well-balanced diet, prolonging the life of the Oscar. Since Oscars are huge waste producers, most owners change their water once or twice every week. Failing to keep nitrates below 40 ppm lowers the Oscar's immune system, and often leads to Hole-In-the-Head disease.

Oscar fish reach sexual maturity at about 14 months when they are 6 to 10 inches long. Breeding Oscars is slightly harder than most other freshwater varieties. The most difficult part is pairing of two Oscars that like enough to spawn with each other. Oscar pairs chosen for mating should be at least two years of age. Size is not reliable, and even age cannot be counted on for getting results. Many mature oscars refuse to spawn until they are 2 or even 3 years old. The only certain way to sex oscar fish is by looking at the breeding tubes. In the female, the tube (ovipositor) is short, stubby and flat at the tip as though sliced off, and about as long as it is wide. The male's breeding tube is about a third the mass of the female's tube, very thin, curved and comes to a point; its length is far greater than its width. The tubes are only apparent just before spawning.

Fig. 99: A pair of Oscars

A 125 or 150-gallon of water would be ideal for these fish and is large enough to house two of them. The Oscar pair should show signs of mating by following a distinct change in normal swimming habits. The pair will begin to slap each other's tails against one another, chase each other around the tank and 'lip lock' mouths. This is where it is good to make sure the Oscars are well behaved and of similar sizes. Unfortunately, at this point one may become aggressive and attack the other.

When this behaviour is displayed between the Oscar pair, we need to give the pair some spawning medium. An excellent choice is offering an upturned dinner plate to the aquarium. The ceramic surface is suitable for the pair to lay and care for their spawn. They will clean the surface in preparation to spawning.

The female Oscar will give birth typically 2-3 days after the mating ritual. The female will give birth in batches of about 100 eggs and typically lay about 1000 eggs in total over the period of a few days. Both parents will vigorously guard the eggs from potential predators. A frequent checking on them too regularly should be avoided, as they may feel threatened and eat the eggs/fry. The female will fan the eggs to prevent fungus growth, as well as remove the unfertilized eggs by eating them. In the wild, Oscar eggs often have a lower level of success rate and even fewer make it to young adults. In captivity, hatching rates and conversion rates can be much higher. The Oscar fry will hatch from the eggs within 72 hours of being laid. It is not uncommon to find all the eggs have disappeared. This can happen due to the eggs not being fertilized and the parents consume them. If the Oscar pair feels unsafe or becomes upset, they may also consume their eggs. It is no surprise that at time parents can take a number of attempts to get spawning right.

Oscar eggs may develop a cotton wool type fungus. This phenomenon will only occur on unfertilized eggs and unfortunately can spread to the fertilized stock too. Adding Methylene blue to the aquarium can remove the fungus but may cause your tank to re-cycle and is best avoided. The parenting pair will usually pick up on the fungus and consume this portion of the stock. Oscar eggs that while are fertilized will be a light brown or tan colour unfertilized eggs will be white in colour

Fry tanks, especially for waste producers like Oscars, are typically bare bottom for ease of cleaning. They need a heater to keep the water temperature between 77 and 80 degrees and a sponge filter to keep the tank cycled. The sponge filter is the safest filter for fry, as other filters such as hang-on-back and canister filters can suck fry into the filter and grind them up in the impeller or coarse sponge. Sponge filters also keep particles of food that the fish miss within reach, so you will probably observe several fry pecking at the sponge filter 24/

7. Fry will die if there is any ammonia or nitrite present, and they are more susceptible to nitrate poisoning than the adults, so the nitrates should never go over 10 ppm in the fry tank. Aside from this, the parameters of the adult and fry tanks should be identical.

IX. Breeding of Flower Horn

Flowerhorn is a type of ornamental fish having bright body colors and flower-shaped head, from where it has acquired its name. It belongs to Cichlidae family of the fishes. Their head protuberance, or *kok*, is formally termed a nuchal hump. Like blood parrot cichlids, they are man-made hybrids that exist in the wild only because of their release. It was first developed in Malaysia, Thailand, and Taiwan. By 1994, red devil cichlids (typically *Amphilophuslabiatus*) and trimac cichlids (*A. trimaculatus*) were imported from Central America to Malaysia and the hybrid blood parrot cichlid were imported from Taiwan to Malaysia and bred these fish together, marking the birth of the flowerhorn. The original flowerhorn hybrid stock are referred to as luohans (from the Chinese word for the Buddhist concept of *arhat*). The four main derived varieties are Zhen Zhu, Golden Monkey, Kamfa, and the golden base group. Flowerhorns have been criticized by cichlid hobbyists and environmentalists for a number of reasons.

Fig. 100: Flowerhorn

The Flowerhorn fish can be bred all through the year as there is no definite time for their mating. Flowerhorn cichlids have a lifespan of 10–12 years. They are usually kept at a water temperature of 80–85 °F, and a pH of 7.4–8.0. They require a tank of a minimum 40 gallons, with 75 gallons being optimal. A breeding pair may require a tank of 150 gallons or more, depending on size. There are several ways by which breeders distinguish between male and female flowerhorns. Generally, the males are larger than the females, but there are some exceptions. Males have the kok, or the nuchal hump, on their foreheads. Males also usually have brighter and more vivid colors. For most breeds, the females have black dots on their dorsal fins, whereas males usually have longer anal and dorsal fins. Females tend to have an orange belly, especially when ready to breed. The mouth of the male is thicker and more pronounced than the females. One sure way to determine the sex of flowerhorn is that grown female will lay eggs every month even without the male. Being aggressive and territorial, two or more flowerhorns are usually not kept together, but the tank housing them can be divided up with acrylic dividers or egg crates. The parent Flowerhorn fishes need to be fed well before leaving them together for mating, as they eat very little during the mating period. Usually, a balanced diet of living shrimps and pellets of readymade fish foods are provided to the adult fishes so that they can produce healthy eggs.

Fig. 101: Breeding tanks for flowerhorn

After selection of the parents of new Flowerhorn fish as per choice of the traits to be seen in the baby fish, these parents should be kept separately by putting a glass divider between them in a large aquarium. They should be well fed to make them healthier before mating. The temperature of the water should be maintained at an optimum level and pH of the water should be neutral. Air should be supplied to the aquarium with an air stone and many decorations should be introduced to that large-sized aquarium. A plate with plain surface should be kept in the aquarium so that the female fish can lay eggs on it. The water of the aquarium needs to be treated with an anti-fungal solution before lifting up the glass divider. After the two fish mates, the female fish lay eggs either over the given plate or on any smoother surface. Then the male fish spread its milt over those eggs to fertilize them. After a few days, the eggs will hatch into tiny Flowerhorn fishes that are termed as fries. Then the parents are taken away and these fries are given special care for proper growth.

After the eggs are laid by the female fish, the filtration may cause damage to these tiny eggs. So, it is better to place an air stone instead of filter for keeping a steady supply of oxygen in the water of the aquarium. After the eggs are hatched, the tiny spawns are removed from their parents and fed with live baby brine shrimps. As these fries of Flowerhorn grow very fast, they need to be fed almost continuously throughout the day. So, it is essential to add food to them around 10 times a day, to ensure that they grow with proper health. When these fries are about 2-3 weeks old, they should be given frozen *Daphnia* as baby shrimps may not be sufficient for them anymore. The Flowerhorn fries of more than one-month-old are given cichlid pellets for keeping them healthy. When the newly bred Flowerhorn fries turn one month old, it is seen that some fries become larger in size than the others and eat most of the food added to the aquarium, leaving the smaller one's starving. So, it is best to separate these differently sized fries to let the smaller ones achieve a healthy growth.

X. Breeding of Discuss (*Symphysodon*)

Discus is colorful, graceful and charismatic tank inhabitants. Native to regions in South America, the discus fish is highly temperamental when it comes to tank conditions. There are three sub-species distinguished by their colors in the wild. Discus are considered to require a higher level of care than other aquarium fish. It is nearly impossible to distinguish male Discus from females. Therefore, it is advisable to ensure a large number of discus fish in order to have a higher probability of having both males and females. The minimum number to give you the best of a breeding pair is at least 12 discus fish. Females reach maturity and start mating at about 9 months of age but males take up to 13 months. The discus must form pairs on their own. It is almost impossible to get a proven male and female and have them breed successfully. As spawning time approaches, the female's abdomen becomes slightly enlarged because of the eggs she is carrying.

Fig. 102: A Discuss

While in the presence of other fish, discus fish become highly territorial during the process of pair bonding. The male discus fish manifests aggressive behavior, usually chasing other fish away from the female discus or the area where they are located. The male discus uses the aggressive behavior to claim the female as a mate and deter other potential mates from approaching the mating pair. It is better to give their pair bonding discus fish a separate aquarium, providing a more relaxed setting for the process. If discus fish are alone in an aquarium, they will begin cleaning the area where they have chosen to spawn. The indicators of this are the fish picking at the area with their mouths, moving rocks, pebbles or anything else that is in the location. Sometimes they clear the site to form a circular area that the fish will use to spawn. The fish do this to ensure the area is safe to begin the spawning process. During the pair bonding process, discus fish often display colors that are more vibrant, which they use to attract their mate. Discus fish also use color display to inform potential mates that they are ready to begin the mating process. A few days before she lays her eggs, the female develops a short tube, or ovipositor, just in front of her anal fin.

The Discus fish will not breed effectively if the aquarium is small. The discus tank should be at least 38 cm deep. This height will ensure the fish tank holds enough water needed by the discus in it. Discus are usually combined with planted aquariums. Availability of nitrites, nitrates and ammonia is of the optimum level. The water pH should be stable at 6.5 whereas temperature should be 86°F. Also, mineral content should be 100-200 microsiemens. These conditions should always be adjusted when they fluctuate in order to achieve stability which the discus fish need for survival and breeding. A change of 10% of the water daily is required for the successful breeding. They must be on animal protein. These come in different forms like black worms, live brine shrimp etc. These will provide the discus with nutrition necessary for breeding. Additional vitamin supplements may help sometimes as well.

Fig. 103: Upturned clay pot for egg laying

Various spawning sites have to be provided in the tank for spawning. Placing an upturned clay pot or cone in the discus tank will give a hard surface for the discus to lay eggs on.

It must be ensured that the surface is clean and will not pollute the aquarium. Professional plastic cones are sold on line that have been used with proven results. These may be a good consideration to increase laying chances.

In most respects, Discus fish spawn like angel fish. They lay their eggs on a vertical surface. The female takes a pass or two before beginning to lay the eggs. In the days before spawning, the couple chose a spot on the wall of the tank, in the substrate or even in the foam of the filter, and start cleaning it with repeated "mouthing's". This behaviour varies in duration from 1 h to 3 days. As spawning approach, is body trembling behaviour can be observed in both individuals. This behaviour characterized the period prior to spawning. Another evidence of this period is the visible prolapse of the urogenital papilla. A few minutes before the beginning of spawning the couple conduct a "spawning rehearsal", in which they perform the entire characteristic ritual of spawning with the inclined body and belly in contact with the substrate, but with no release of oocytes.

Once she starts truly laying the eggs, the male will come right behind and deposit the sperm.This continues for several passes, and with each one, more eggs are deposited and fertilized. The eggs hatch in about 3 or 4 days, during which time the parents guard the eggs, fan them with their fins, and work them over with their mouths, cleaning them of fungus or any foreign objects that may befall them. As the larvae develop, they begin swimming in an errant way, falling off the substrate. The parents collect the larvae in their mouths and return them to the substrate. At this stage, the parents looks darker in their color patterns, which attract the offspring to their parents because the discus larvae at the beginning of orient swimming exhibit negative phototaxis, in other words, they get attracted to darker objects.Parents are absolutely necessary to the development of the fry. Removal of the parents from the fry becomes inevitable to starve and kill the fry. Research says that there is special food producing cells or glands in the parents' skins. As the babies reach the free-swimming age, they cling to the side of one parent, feeding while they cling. When one parent grows weary of the babies, it shakes itself, and all the babies are transferred to the side of the other parent. For several weeks the fry continues receiving nourishment from the parents.At first, the tiny, sliver-like fry remain nearly motionless on the spawning site, unless the parents move them to a new location, which they seem to do quite often.

Fig. 104: After spawning, the couple showing parental care of the eggs

A young Discus bears little resemblance to their parents. For the first months, they are elongated, like most cichlids. But by the time the body of the fish reaches the size of a dime, they are nearly as rounded as the adult.

8

Breeding and Seed Production of Indigenous Ornamental Fish

The indigenous ornamental fish which are very popular overseas, are being harvested from the wild and exported. This in the long run, may lead to depletion of stock due to over harvesting of our valued native fish species. Hence, breeding of native fish species is also equally important for a sustainable trade of ornamental fishery and to conserve the natural stock. Some fishes breed naturally with ease, whereas others are bred by artificial techniques like hand stripping and hypophysation. In nature, the fishes are stimulated by changes in environmental factors. These conditions can be simulated to some extent in the aquaria by providing specific biological requirements for spawning.

I. Breeding of Zebra danio (*Brachydanio rerio)*

Brachydanio rerio is commonly known as zebra danio. The zebra danio, as the name implies, is characterized by a number of bright blue or purple horizontal lines, which run from the head to the caudal fin. This pattern is repeated in the anal and caudal fins too. The flanks have prussian blue colouration and operculum has golden blotches. The dorsal fin of this species has an olive yellow base and there are blue gold markings on the anal fin.

Determination of sex

The females were found to be more silvery and larger in size when compared to males. The belly is swollen in females, especially in gravid ones; while in males the body is generally slim.

Spawning

The zebra danio is a prolific breeder and their breeding season commences from April and continues till August. The eggs are non-adhesive and their breeding habit is of egg scatter type. The species can easily breed in aquarium tank. Medium size all glass aquarium tanks (60 cm×30 cm×30 cm) are suitable for breeding. Aquarium tank, provided with a perforated tank divider, was found to be suitable for the conditioning of the zebra danio. The males are introduced on one side and the females on the other side of the divider, in the same aquarium tank. The fishes when conditioned on a good variety of diets including live feed for about two weeks were ready for breeding. Successful spawning was recorded by introducing a shoal of 20 fully matured vigorous males and females in a ratio of 3 males: 2 females into the spawning tank. The spawning tank contained about 10 cm of water with a steady temperature of 26-29°C, and gentle aeration. A water pH of around 7.0 and hardness of normal to 250 ppm has given good results. Bottom of the aquarium should be completely covered with pebbles of about 6 to 8 mm diameter. After spawning, the eggs fall in between the pebbles, arranged at the bottom of the breeding tank. This prevents the eating of eggs from cannibal male. A net fixed along the bottom of the tank (at a height of 5 cm from the bottom) instead of pebbles are also found to be successful in protecting the eggs from the parent fishes. Both male and female should be removed after spawning from the tank.

Hatching and larval rearing

The eggs require a hatching time of 2 days, if temperature is favorable. As soon as the tiny hatchlings are seen in the aquarium tank, the parents should be removed. The hatchlings takes 2 days to absorb their yolk sac. After 2 days, they should be fed with infusorians for 4 days. Subsequently, rotifers, smaller daphnia can be fed for a week. After that they start feeding on larger plankton and powdered formulated feed.

II. Breeding of Moustache Danio (*Danio dangila*)

Danio dangila is a beautiful ornamental fish species which attains a length of about 15 cm (15 g) suited for aquarium keeping. They are distributed in Arunachal Pradesh, Assam, Bihar, Meghalaya and West Bengal. This species is known from mountain streams of Ganga-Brahmaputra drainage.

Determination of sex

Males are small, bright with slender and ventrally straight belly, slightly rough with red linings on anal fin and ovor anal openings. Females are larger, slightly dull with bulgy and ventrally curved belly, soft anal fin with yellow lining and round anal opening.

Spawning

For induced breeding at the ratio of 2:1 (male:female) with average weight of mature male is 4.5-5.5 g (4-5 cm) whereas female is 6.5-7.5 g (6-7 cm). Hapas, prepared of 12 X 12 x 12 cm dimensions, can be fixed in the aquarium tank of size 30 X 60 X 45 cm keeping the water level at 20 cm height. Water parameters should be maintained at optimum; temparture, 20°C, pH: 7.1, DO: >5.2 ppm, hardness: 30-30 ppm and alkalinity: 40-45 ppm. Fishes are injected at pectoral fin base with WOVA-FH @ 0.5 ml/Kg for female and @ 0.3 ml/Kg for males. Spawning occurs after 16 hrs of injection with 1100-1200 eggs and 85-87% hatching rate. After spawning, hapa with parents should be removed from the tank.

Larval rearing

The newly hatched larva is transparent and yolk sac is absorbed after 70-72 h of hatching.

III. Breeding of Silver Danio (*Danio devario)*

Body looks silvery, and olivaceous green on the back. A dark longitudinal band in the middle of sides extending from caudal fin base to below the dorsal fin origin. The Silver danio is found in lower region of hill stream, plain rivers and lentic wetlands like beels with clear slow running water. This species usually swims on the lower surface of upper column zone in shoal.

Determination of sex

Body shape slender in male but it is stumpy in female. Dorsal profile is slightly convex in male. In female, both dorsal and ventral profile are distinctly convex. In male, the ventral edge is prominently keeled whereas in female ventral edge is comparatively less keeled. Mouth is blunt in male but in female it is pointed. Dorsal and anal fins are large in male but fins are comparatively small in female. Body colour bright in male and in female it is faint. In male the dorsal, caudal and anal fins are reddish yellow in breeding season. Fins are pale yellowish in colour during breeding season in female. In natural population, the sex ratio is 2:1 (Male: Female).

Spawning

The breeding season is recorded in the month of May to August, but the peak occurs in June. In nature generally the species select a suitable spawning ground, where uprooted tree and broken tree branches are submerged. The species can easily breed in aquarium tank. Medium size all glass aquarium tanks (60 × 30 × 30 cm) are suitable for breeding. Spawning ground is to be prepared with smooth glass bead, placed at the bottom of aquarium tank in 3-4 layers or mixture of sand and gravel with a minimum thickness of 3-4 cm. Coarse vegetation in the spawning bed is congenial for breeding. The fully gravid female is introduced in the breeding tank in morning hours. The male fish is introduced in the tank after 10-12 hours of female introduction. After completion of courtship, the female releases the semi-adhesive eggs in late night and completes the spawning activity in the early morning. The eggs are scatterly released and fall inside the gap of glass bead, which protects the eggs from cannibal male. Both male and female should be removed after spawning from the tank.

Hatching and larval rearing

The eggs require a hatching time of 2 days, if temperature is 24-26°C. As soon as the tiny hatchlings are seen in the aquarium tank, the dead eggs and other dirt should be removed. The hatchlings takes 2 days to absorb their yolk sac. After 2 days, they are fed with infusorians for 3 days. Subsequently, rotifers, and smaller daphnia are fed for a week. After that, they start feeding larger plankton and powdered formulated feed. The hatchling number is estimated between 250-300 numbers per female.

IV. Breeding of giant danio (*Devario aequipinnatus)*

The giant danio, *D. aequipinnatus,* is an active and brilliantly striped ornamental hill stream fish and has a great value in the ornamental fish market. It inhabits hill streams up to an elevation of 300 m and does not grow more than 15 cm. It is found in shaded, mid-hill clear waters with pebble or gravel substrates and occurs in schools at the surface in small high-gradient upland streams.

Determination of sex

The mature females are larger when compared to males and have a rounded belly while the males are slender and streamlined and more colourful.

Spawning

Big size glass aquarium tanks (1.5 x 0.5 x 0.5 m) with dark substrate (marbles) and natural day lighting provided with spawning mops made of green wool is ideal for spwaning. In all the tanks for rearing and breeding experiments, pH is maintained in the range 6.7-7.0, nitrates, nitrites and ammonia should be maintained at 0, hardness in the range of 60-70 ppm, total alkalinity in the range of 30-35 ppm and dissolved oxygen in the range of 7.5-8.5 ppm. In the spawning tank, one gravid female can be released along with two mature males. If a pair is compatible then the male will constantly chase the female all around the tank and hit the female in its abdomen with its head for spawning. After this activity of 20-25 min, they start spawning after the sunrise. The female releases eggs in batches of 15-20 and the eggs are immediately fertilized by the male sideways and they continue spawning frequently up to 6-7 days till the female releases all the ripe eggs. The female in the first spawning batch releases 50-60 eggs all around the aquarium. The percentage of fertilization in this case was seen to be 75-80. Larvae hatches after 34-36 hrs after fertilization.

Larval rearing

On sixth day, larvae start swimming and feeding with infusorians as starter feed after that shifting to zooplankton.

V. Breeding of Rasbora (*Rasbora daniconius*)

Rasboras are considered as egg depositor, laying eggs in flat leaf or substratum. They are the most important members of the egg layer group and show the highest degree of parental care. The very common species is *Rasbora daniconius*. The body colour is greenish yellow with silvery sides. Fins are pale orange and the caudal lobes are tinged with black.

Spawning

The species can easily breed in aquarium tank. Small size all glass aquarium tanks (30×15×20 cm) are suitable for breeding. Before spawning, they select the site to lay eggs. The chosen site may be filter tube, glass wall or plant leaf. They will forcibly evict any other fish from the surrounding area by chasing them away. Generally, 1 female and 2 males are released into breeding tank. The breeding tank is provided with flat leaved plant or flat substratum for depositing their eggs. Generally, one female lays about 100-250 eggs.

Hatching and larval rearing

The eggs laid on the underside of the flat leaves will hatch after 24-30 hours and hatchling becomes free swimming after 2-4 days. At this stage, the tiny hatchlings should be fed with infusoria and newly hatched brine shrimp. As they grow bigger, they should be fed with zooplankton like *Moina* and *Daphnia*.

VI. Breeding of Rosy barb (*Pethia conchonius*)

The *Pethia conchonius,* popularly known as Rosy barb, is an attractive ornamental fish. The maximum length of the fish is about 14 cms. During breeding season, fish becomes more colorful. The back of the body is a shiny-olive green whereas the rest of the body is mainly silvery and tinged with soft pink color, from which its common name was derived.

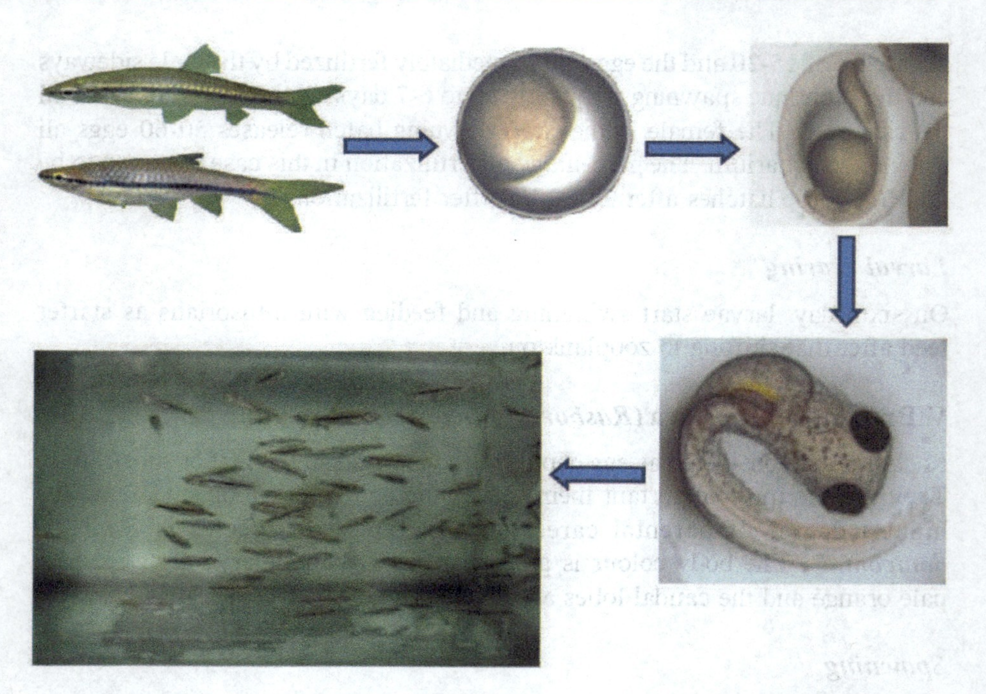

Fig. 105: Captive breeding and rearing of Rasbora

Determination of sex

The fish sexually mature when they reach 6 cm. Both sexes are generally glossy green and silver with a black spot near the tail and, with black tipped fins. In ripe condition, the male blushes a brilliant rosy pink color while the fin becomes red, with dense black tip. During the breeding season, the females develop a swollen abdomen.

Fig. 106: *Pethia conchonius*

Spawning

Barbs perform a 'Spawning run' as part of their courtship ritual, and so they must have a tank that is sufficiently large to allow them to gather enough speed for their runs. It is required to provide thin layer of gravel, clumps of plants for the eggs to scatter into. The males are aggressive during the spawning period and in small aquarium may thrash unripe females to death. It doesn't happen if the fish are in shoal and in a large aquarium. The water should be well aged and have a neutral pH. One female and two males can be placed in the tank. Spawning usually takes place in the morning and the females are more active partners. They chase each other around and the spawning takes place in one of the plants placed in the aquarium. The pair wraps themselves around each other and shakes until the eggs are laid. It happens several times and the number of eggs are very large. The fry hatch in 24 to 48 hours at a temperature range of 26-30°C. The parents are egg eaters and also eat the spawn. Removal of the trio immediately recommended as they will start eating the eggs even during the spawning.

Fig. 107: Breeding set up in tanks for Rosy barb (*Pethia conchonius*)

Fig. 108: Eggs of Rosy barb attached to the leaves

Larval rearing

The hatchlings are fed with infusorians after the absorption of yolk sac for about 7 days and thereafter should be given sieved small zooplankton and chopped tubifex worms.

VII. Breeding of Pool Barb (*Puntius sophore)*

The Pool Barb, *Puntius sophore,* is a potential indigenous ornamental fish and popular among the hobbyists all over the world due to their attractive coloration, graceful and majestic movements. The breeding season is during May to July.

Determination of sex

The fish sexually mature when they reach the age of 7+ months in captivity. The minimal length and weight at first maturity of female fish were 6.5 cm and 3.2 g, respectively whereas male attained maturity with a length of 6.0 cm and weight of 2.5 g. Males are generally small, bright and identified on the basis of slender and ventrally straight belly, slightly rough and oval anal opening. Females are larger, slightly dull with bulgy and ventrally curved belly and round anal opening. Fins hyaline in mature females; anal and pelvic fins brick red in mature males.

Spawning

Induced natural breeding is succeded by administering intra-mascular injecton with Gonopro-FH hormone of a single dose @1 ml/kg body weight to matured female and @ 0.5 ml/kg body weight to matured male. The fish spawned after 6 to 7 hours of administration of inducing agent at 27-30^{0}C. The spawning can be observed by the presence of eggs on the spawning mops and bottom of the aquarium. After the spawning, parents should be transfered to another aquarium; this prevents the eating of eggs by the parent fishes.

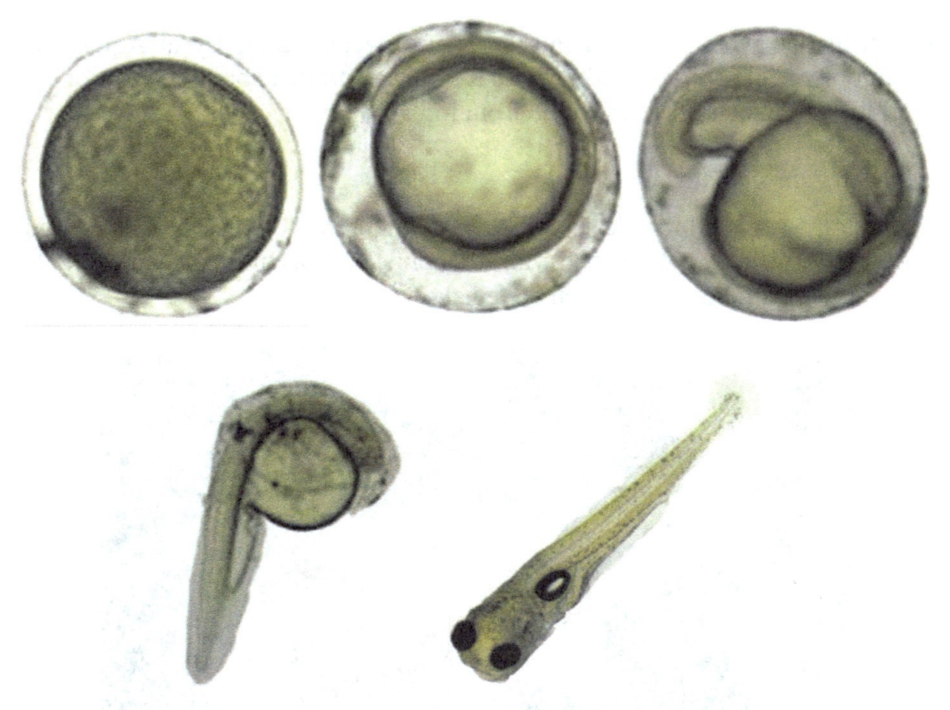

Fig. 109: Breeding of *Puntius sophore*

Due to the slightly adhesive nature of the egg, considerable debris adheres to the capsule of the egg. The number of eggs from a single spawning act varies from 2500 to 3500. Fertilization and hatching were 90% and 70%, respectively.

Larval rearing

The newly deposited fertilized eggs are demersal, slightly sticky in nature, translucent, un-pigmented, light yellowish in colour and almost spherical in shape with a diameter ranging from 990-200 μm x 870-600 μm. The seven broad period of embryogenesis-zygote, cleavage, blastula, glastula, segmentation, pharyngula and hatching have been identified and the duration of each period is recorded. Hatching takes place after 14-16 h of incubation. The hatchlings are transparent, 3.0±1mm in length with a large oval head, well defined large yolk sac and short tail.

VIII. Breeding of Miss Kerala (*Puntius denisonii*)

Puntius denisonii is endemic to Western ghats where it occurs in fragmented populations in states of Kerala and Karnataka. *P. denisonii* is much sought after fish in the international ornamental fish trade. It received fame after this beautiful fish won an award in the "New Species" category at Aquarama exhibition of Singapore in 1977. Miss Kerala constituted about 60-65% of the total live ornamental fish, exported from India.

Determination of sex

The mature male and female should be separated based on their coloration and body depth; brightly colored fish with streamline body as male and slightly pale colored fish with bulgy abdomen as female.

Fig. 110: *Puntius denisonii*

Induced spawning

For induced breeding male with brighter color (8-9 cm) and female with bulgy abdomen (10-12 cm) are selected with 2:1 ratio (male : female). As inducing agent, a small dose of WOVA-FH or OVATIDE hormone (0.7 ml/Kg) can be applied. Male injected with half dose of female. Male and female are released into the breeding tank (Temp: 22-26°C, pH: 6.8-7.0, TDS <30mg/L). After 16-18 hrs of injection male and female are collected from breeding tank for stripping. Strip female on clean dry watch glass and then strip the male so that milt drops over the eggs in the watch glass. Mix the eggs and milt slowly, using fine brush or feather. Wash eggs with distilled water and remove the excess milt from the watch glass. After repeated washing, fertilized eggs are placed in hatching tank (Temp: 24-26°C, TDS <50 ppm).

Larval rearing

Larvae hatch after 34-36 hrs of fertilization. Yolk sac absorption takes place on 3rd day after hatching. After 3 days, larvae are fed with micro worms. As they grow bigger, they should be fed with zooplankton like Rotifers, Moina and Daphnia till 16th day and after 16th day Cyclopes/Copepodes. From 22nd day onwards they are fed with formulated feed.

IX. Breeding of Dwarf Gourami (*Colisa lalia)*

The *Colisa lalia* is the most popular of the gouramis. It is a native to North East India. It lives in freshwater ponds, streams, ditches and paddy fields. The original natural dwarf gouramy is reddish type. It is stocky and laterally compressed with an ovate shape when viewed from the side.

Determination of sex

The species exhibits striking sexual dimorphism, with males showing exquisite colours while females have a silvery colouration. The male is slightly larger than the female and grows up to 6 cm. The dorsal and anal fins are more developed in the male. The pair of ventral fins is filamentous and almost as long as the body.

Spawning

The breeding season is recorded in the month of May to August, but the maximum predominance occurs in the month of June. The species can easily bred in aquarium tank. Small size all glass aquarium tanks (30 cm×15 cm×20 cm) are suitable for breeding. A piece of dried banana leaf or a piece of thermo coal are put to float on the surface of the water in breeding tank, for the male to build a bubble nest on lower side of leaf/thermocol. The breeding containers are covered completely with canvas sheet. During and after making the nest, the male displays to the female which usually ends with both the fishes embracing the nest resulting in deposition of large number of eggs in the nest. The female lays about 1000-1200 eggs per spawn. After breeding, the female is removed as the male guard the eggs that remain attached to the floating bubble nest. The male picks up the eggs with its mouth and place them in the bubble nest and looks after the bubble nest and eggs.

Hatching and larval rearing

The eggs hatch in about 24-26 hours after fertilization and become free swimming in 3 days. The moment the fry begin leaving the nest, the male also should be removed from the tank. The free-swimming hatchlings feed on infusoria as a starter feed. After absorption of yolk sac for about 10 days, the fry start taking newly hatched *Artemia* and small cladocerans. During this stage, fry require vigorous feeding and is given sieved small zooplankton and chopped tubifex worms. Subsequently, when they grow little bigger, they are stocked in bigger cement tanks for further growth.

X. Breeding of Peacock Eel (*Macrognathus aculeatus)*

The spiny eel, *Macrognathus aculeatus,* is commonly known as porthole eel or the peacock eel. The body is long and eel-like with a long fleshy snout and a rounded tail fin that is separated from dorsal and anal fins. The body colour is brownish to yellowish ventrally and marked with two long dark bands on either side. There are 3-11 ocellii (false eye spots) at the base of dorsal fin. Both the dorsal and caudal fins have several fine streaks. Generally, it reaches a maximum size of 25 cm weighing 60 g. It is in high demand as ornamental fish in the export market due to its beautiful body shape, coloration and playful behavior. Peacock eels seem to have nocturnal feeding habits, preferring to hide by burying themselves in the substrate or under rocks during the day. They come out at night and early morning to feed. Since they prefer to hide, it is better to provide shelters such as pieces of bamboo, submerged aquatic weeds and pebbles to create a congenial environment for breeding.

Fig. 111: Peacock eel

Determination of sex

It is difficult to determine the sex of the fish when they are young. In general, the females are slightly larger than males of the same age. During the breeding season, the females develop a swollen abdomen with a greenish tinge, while the males ooze milt when gentle pressure is applied to their abdomen.

Induced spawning

The breeding season starts with the onset of monsoon in the month of April and lasts till August with peak breeding in June-July. For captive breeding, the breeding tank needs to be provided at least 4 cm thick layer of small marble stones and some water hyacinth as substrate for spawning. As inducing agent, a small dose of Ovaprim hormone (0.025-0.05 ml per fish) can be applied. Breeding can be done in small glass aquarium with a water depth of 10 cm. Mild flow of water is conducive for breeding which can be maintained with the help of an electrical filter. Spawning response varies from 8-10 hours. Courtship begins with the male chasing the female and swimming in a tight circle. Later, the pair encircles each other around the water hyacinth for spawning. The eggs stick to the roots of the water hyacinth. The male releases sperm as the eggs are laid. The eggs are round, green in colour, and adhesive in nature and about 1.25 mm in diameter.

Larval rearing

The yolk is fully absorbed in 4 days. The free-swimming fry hide among pebbles, plants and in other substrate. Larval feeding is very crucial. Once the yolk is absorbed, it needs to be fed with infusorians, zooplankton and boiled egg yolk. They readily accept tubifex worms. This fish can be reared in community aquarium tank as they are generally peaceful in nature.

XI. Breeding of Frail Gourami (*Ctenops nobilis)*

*Ctenops nobilis*is a species of gourami, native to northeastern India and Bangladesh, found in lakes, ponds, rivers, and streams with plentiful vegetation overgrowth and regarded as one of the most difficult fish to keep. It is a mouthbrooder fish. A very soft, acidic water (pH 4-5) is usually recommended for this species. The ideal temperature for fish survival and growth is around 20-32°C. The fish is generally slow moving and shy in nature so they usually prefer to relax under the aquatic vegetation like, *Vallesnaria* sp., *Hydrilla* sp. etc. The fish is mainly carnivorous in nature and their preferable food items varies between zooplankton, micro worms, tiny crustaceans, insect larvae etc. In captive condition, Tubifex and mosquito larvae give the best gonadal maturation. The adult male fish can be clearly recognized with their broad "spouts" and curved dorsal fin base than the female. Females mainly attain first maturity at the length of 75-80 mm and weight of 5.10-6.59 g and males attain first maturity at 70-74mm length and 5.05-5.78 g weight. The GSI attains highest peak in August and the lowest has been observed in the month of December. The breeding season of the fish extends from late July to September. In natural habitat, the sex ration of the fish is approx. 2:1 (Male: Female). The male fish chase a selected female from a sole. The mating process continues up to 3-4 days. The breeding process takes several hours to complete fully. Mouth breeding parental care is observed in both male and female fish. At the time of carrier, they cannot accept any food. After 9-10 days of spawning, the carrier fish start to spat out 10-15 numbers of free swimming larvae at the size ranging between 3.5 and 6 mm and the process continued up to 18-20 days. A total of about 110-165 larvae can be produced from one pair of fish. At this time, the water temperature should vary between 28-32°C, pH should be 7.5-8. Any external stress for the fish can lead to engulf the eggs. The first food of the fry is Daphnia up to 4-5 days.

Mature fish Mating Mouth Brooder female

Mouth brooder male

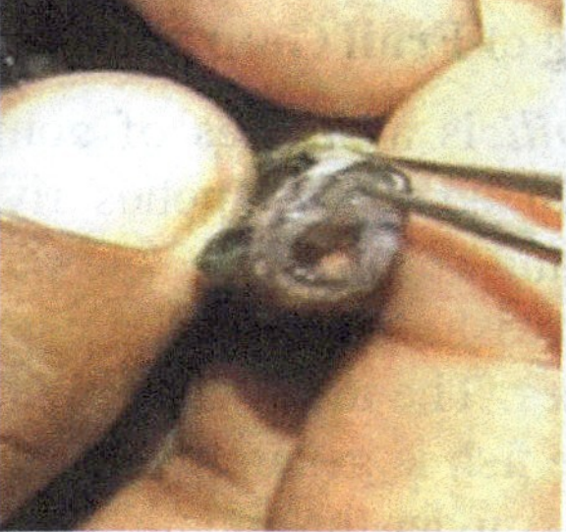

Egg in mouth

Free swimming larvae

Fig. 112: Breeding of frail gourami fish

9

Genetic Improvement of Ornamental Fish

As the ornamental fish industry expands throughout the world, genetic considerations have become more important in terms of production, sustainability and conservation. Genetic technologies are now being applied to increase food production in aquaculture through traditional and modern breeding programmes. Genetic markers can help to manage capture fisheries and distinguish farmed from wild stocks in mixed fisheries. Genetic technologies are also now finding application in post-harvest and trade to assist in product identification and traceability. Genetic improvement of fish has considerably increased the fish production from aquaculture. It includes selective breeding, hybridization, polyploidy, gene and nuclear engineering. For any genetic selection programme, induced breeding is the basic and initial step. Intensive selective breeding is practiced with various species of fish in perfecting appearance, shape and color so that new strains are produced.

Domestication

Most of the fish being widely grown in captivity are domesticated to some extent, but there are very few which have undergone genetic improvement. Just as terrestrial livestock and crops are bred to be not only productive, but productive in convenient proximity to humans, aquatic plants and animals are now undergoing these same pressures.

In comparison to agriculture crops, fruit plants and animals, aquaculture is much behind with regard to the domestication of the fish, especially ornamental fish. Never the less, a good number of ornamental fish have been domesticated which can be considered for selective breeding to improve their traits. The available technologies and tools could be used to remarkably speed up domestication and genetic improvement process.

Domestication implies much more than merely keeping wild animals in farms or homes. Domestication is a very long process, during which captive individuals become more adapted to captive environments. Domestication leads to permanent genetic modifications of a bred lin-eage, while taming or keeping wild fish in captive conditions is only conditioned behavioral modification of individuals. The entire life cycle of the domesticated fish species must be fully closed in captivity. Once full life cycle is completed in captivity, the process of domestication can proceed further.

The first truly domesticated fish species was certainly the goldfish, for which the domestication was initiated by the Chinese about 1500-2000 years ago.

Genetic markers

A genetic marker is a gene or DNA sequence with a known location on a chromosome that can be used to identify individuals or species. It can be described as a variation (which may arise due to mutation or alteration in the genomic loci) that can be observed. Genetic markers are useful for determining parentage because offspring share markers with their parents. They are also expected to be useful for identifying regions of DNA that confer performance advantages.

If a particular marker is always found associated with resistance to a cold-water bacterial disease, for example, selection for breeding could be based on the presence of this marker without doing a challenge trial under tightly controlled environmental conditions. This application of selective breeding is referred to as marker-assisted selection.

Males and females are often externally indistinguishable in many fish species, resulting in cases where two animals of the one sex have been placed together in unsuccessful attempts to breed. Genetic markers on the sex chromosomes can be used to distinguish the sexes in fish species. Various molecular markers, protein or DNA (mt-DNA or nuclear DNA such as microsatellites, SNP or RAPD) are now being used in fisheries and aquaculture. These markers provide various scientific observations which have importance in ornamental fisheries. These can be used in: 1) Species identification 2) Genetic variation and population structure study in natural populations 3) Comparison between wild and hatchery populations 4) Assessment of demographic bottleneck in natural population 5) Propagation assisted rehabilitation programmes. DNA Barcoding technique can be used in monitoring the trade of threatened species and demand conservation strategy for sustaining wildlife.

Selective breeding

The basic tool available for genetic improvement, selective breeding, entails choosing the animals with the highest genetic value as breeders for the next generation. To determine this genetic value, populations are raised in a standardized environment so identified differences will be due to differences in genes, not the environment. Selective breeding is a tool used for improving specific traits that provides benefit to the environment as well as to business. Traits like growth performance, disease resistance, pigmentation etc. can be addressed through this process.

Directional filtering of traits is an important and indispensable step in genetic breeding programs. Recently, research has focused on methods to rapidly and efficiently screen for important economic traits. Advances in the fields of genetics, molecular biology, and other biological techniques have resulted in a shift from single traditional selective breeding to more diverse selective breeding methods. Selective breeding can be attempted through following methods:

- Traditional selective breeding
- Molecular marker-assisted breeding
- Genome-wide selective breeding technology
- Breeding by controlling single-sex groups

Table 6. Fecundity of different ornamental fish species

Species	Average fecundity	Spawning/ Year
Molly/Guppy/Sword Tail	200	10
Blue Gourami	3500	10
Pearl Gourami	800	10
Rosy Barb	700	10
Tiger Barb	500	10
Zebra/ Pearl/ Veil tail Danio	1000	10
Angel	800	12
Black widow Tetra	3000	10
Serpae Tetra	800	10
Gold fish	3000	3

Source: Jameson and Santhanam (1996)

Cross breeding

The cross breeding (hybridization) of two species is one of the most basic methods of integrating desirable traits (e.g., growth rate, disease resistance). Hybridization aims to get improved new varieties, having better growth rate, reproduction rate, disease resistance, yield and quality through integrating excellent parental characters, which plays an important role in fish breeding improvement and production value. The process of hybridization is divided into two categories, distant and close hybridization, based on the parental genetic relationship. Distant hybridization characterizes crosses between parents that differ by species or higher classifications. Close hybridization characterizes crosses between parents of the same species, but different strains, different varieties, different ecological types, or different populations of individuals. As such guppy (*Poecilia reticulata*) is an interesting aquarium fish and have been the subject of experiments and research by physiologists, geneticists, evolutionists and ethologists mainly in view of the following features : (i) hardy nature, (ii) being a live bearer and easier for breeding and genetic manipulation, (iii) having distinct sexual dimorphism besides extreme varied colorations in males.

Fig. 113: Cross bred of Endler/Guppy

Not all livebearers can hybridize; a Guppy and a Platy for example, cannot cross.

Guppy x Endler - They will have the elongated fins of the Guppy, but the orange/black distinctive spots of the Endler, they will also be smaller than a Guppy. They can vary in pattern greatly, and many hybrids are now been selectively bred to create specific strains, such as the Tiger Endler. Many Endlers bought in pet stores are often Guppy crosses, true Endlers have rounded short fins, the crosses will have more angular longer fins. Female Guppy/Endler hybrids will be harder to spot due to the lack of colour compared to males, it's best to rely on the tail shapes to identify them.

Platy x Swordtial - Closely related, and create fish that look much more different than a Platy. They are known as the sword-platy or the platy-tail.

Platy x Platy - There are several Platy species. Platys are usually hybrids already.

Molly x Molly - (or *Poecilia sphenops*), like Platys molly also can be hybridzed within differen species of molly.

Molly x Guppy - Called a "Muppy" or "Golly". This does happen, but the fry are sometimes very weak and do not live long. Even if they are stronger they are usually non descript in coloration and fins and people rarely persist with them.

Traceability

Traceability is a key aspect of certification and eco-labeling schemes. Genetic markers provide an extremely sensitive means to identify samples of fish, such as frozen material, fillets and early life history stages, for example, eggs and larvae, which are hard or impossible to identify by other means. Molecular genetic diagnoses of fish have already identified cases of mislabeling and consumer fraud, and have helped convict offending parties. Traceability is defined by the Codex Alimentarius Commission as "the ability to follow the movement of a food through specified stage (s) of production, processing and distribution". Traceability facilitates knowledge regarding the identity, history and source of a product, or of materials contained within a product. It also facilitates knowledge regarding the destination of a product, or any ingredient contained within it. Traceability systems are therefore information management tools. In the fishery sector, traceability information is used in relation to: a) food safety: to ensure that products and materials from which they are made, come from origins that meet food safety conditions b) application of tariffs and quota tariffs, to ensure that appropriate rates of duty are applied c) ensuring that the fish is derived from sustainable sources, such as from vessels which follow conservation rules (e.g. for catch certification schemes).

Traceability may be an explicit requirement set out in regulations. These may be national requirements, or be applied as condition of supply to an export market. This is the case in the EU, where there is a specific requirement expressed under Article 18 of Regulation (EC) No 178/2002 of the European Parliament and of the Council of 28 January2002, laying down the general principles and requirements of food law, establishing the European Food Safety Authority, and laying down procedures in matters of food safety. The EU therefore requires all food business operators, feed producers and primary producers of animals to have in place a "one-up and one-down" traceability system. This provision is applied to food and feed business operators in third countries which supply the EU with products, under the requirement that such supply should be subject to conditions at least equivalent to those set out in EU legislation.

The Code of Federal Regulations requires importers to the U.S. to maintain records that identify the immediate sources of their foods. They must maintain these records for at least two years and make them available to the US Food and Drug Administration (USFDA) within four hours, if requested. The Bioterrorism Act of 2002 also requires domestic and foreign facilities that manufacture, process, pack or import food for human consumption in the United States to register with the US FDA.

The trade of ornamental species has experienced a significant expansion world wide; however, this industry still relies on a large number of unsustainable practices (e.g., cyanide fishing, over exploitation of target species) and needs to shift its operations urgently to avoid collapsing. Under this scenario, traceability and certification emerge as important management tools that may help this industry to shift toward sustainability. This industry relies on the trade of thousands of small-sized species that are traded live on a unitary basis with high market value. These features, along with a fragmented and complex supply chain, make the traceability of ornamental species a challenging task. There are many methods, used for tracing aquatic organisms and some of them are suitable to trace ornamental fish species as well. The use of bacterial fingerprints appears to be the most promising method to successfully trace ornamental fish, but it is most likely that a combination of two or more traceability methods need to be implemented to cover all the unique features displayed by the live trade of ornamental fish species.

Transgenesis

Transgenesis refers to the technique of incorporating a gene or genes through biotechnological methods, not breeding. Soybean plants that have been made resistant to the effects of herbicides are an example of transgenic crops. With a gene from bacteria transferred into and expressed in the plants, their use reduces labor costs for cultivating and weeding. Golden rice, a transgenic rice plant, has three foreign genes inserted from bacteria and daffodils to allow the production of vitamin A in the rice.

The Glow fish, a fluorescent zebra-fish which came to the U.S. market in January 2004, is the first transgenic aquatic organism to be marketed. The proteins responsible for making the zebra fish fluorescent are transgenes derived from jellyfish and coral. These fish are not intended for food.

Atlantic salmon with a growth hormone gene from Chinook salmon and a promoter, or switch, from the ocean pout are currently being reviewed by the United States Food and Drug Administration for approval as food fish. Research on other aquatic transgenics with the potential for greater disease resistance, enhanced nutritive value, and other characteristics is ongoing around the world.

Developmental genetics in pigmentation

Vertebrate pigment patterns are both beautiful and fascinating. In mammals and birds, pigment patterns are likely to reflect the spatial regulation of melanocyte physiology, via alteration of the colour-type of the melanin synthesized. In fish, however, pigment patterns predominantly result from positioning of differently colored chromatophores. Theoretically, pigment cell patterning might result from long-range patterning mechanisms, from local environmental cues, or from interactions between neighboring chromatophores. Recent studies in two fish genetic model systems have made progress in understanding pigment pattern formation. In embryos, the limited evidence to date implicates local cues and chromatophore interactions in pigment patterning. In adults, de novo generation of chromatophores and cell-cell interactions between chromatophore types play critical roles in generating striped patterns; orientation of the stripes may well depend upon environmental cues mediated by underlying tissues. Further genetic screens, coupled with the routine characterization of critical gene products, promises a quantitative understanding of how striped patterns are generated in the zebra fish system. Initial 'evo-devo' studies indicate how fish pigment patterns may evolve and will become more complete as the developmental genetics is integrated with theoretical modelling.

Challenges

Genetic improvement is a major contributing factor in developing efficient aquaculture, as it is in other forms of animal production. Moreover, as most fish species are only slightly domesticated, the potential gains are very large. Many of the current methods are being used in livestock rearing (choice of breeds, cross breeding, selective breeding). Few are very specific to aquatic organisms (genetic control of sex, sterilization through triploidization). The spread of selected lines in aquaculture has been effective only since the 1990s, with lines selected essentially on growth and producing some impressive results. Transgenesis, although technically feasible on growth rate, is not used in practice and will become of use only through the definition of new phenotypes, beneficial for all and out of reach of classical selection. In order to allow a sustainable development of ornamental fish industry, selection criteria should evolve towards an improvement of production efficiency (feed efficiency, disease resistance) and adaptation to new contexts (fluctuating environment, vegetal-based feeds). An improvement of selection in such a direction, in addition to the high potential of "classical" methods, should take advantage of the development of genomic technologies, first to make the best use of molecular pedigrees and in the longer term to improve the knowledge in the genetic and biological bases of traits and to select directly on the genotype (genomic selection).

Artificially selected/genetically modified livestock versus conserving natural biodiversity

Fishes, invertebrates and plants in many public aquarium collections often originate from the ornamental trade. The majority of freshwater fishes in the trade are derived from captive breeding and culture, but this presents scientific, conservational and management problems for public aquariums. This is in terms of the lack of a breeding history, possible hybridization, the modification of natural behaviors and the absence of a natural geographical provenance for the livestock. Artificial selection and genetic engineering in ornamental and scientific laboratory facilities to produce 'attractive' or 'useful' traits (actually sometimes 'monstrous' or 'transgenic' forms which would barely survive in the wild) poses further problems. Some transgenic aquarium fishes such as 'medakas' (*Oryzia slatipes*) and zebra fish (*Danio rerio*) incorporate jellyfish genes that cause the fish to glow in the dark; and there is a real prospect of such genetically modified (GM) fishes being released to the wild with uncertain consequences. There are overall concerns in terms of animal welfare and also the ex situ maintenance of representative natural aquatic biodiversity and natural genomes, with the longer-term prospect of re-introduction back to the wild where necessary and appropriate. The widespread cultural practice of die-injection to add 'day-glo' artificial colors to the fishes more subdued natural pattern is an associated welfare problem.

10

Fish Health Management

Ornamental fishes are very prone to diseases due to their captive habitat. Fish health management is an important issue of concern at all the ornamental fish production facility. Proper water quality is required to be maintained in the aquarium for the good health of the fish. However, the chance of infection in these fishes is very high. Many limiting factors viz. water quality, over crowding, nutritionally imbalance food, rapid fluctuation in water temperature, lack of oxygen or poor husbandry practices influence the health status of the stock. The symptoms/signs indicate whether the fish is suffering from a particular disease or not. If the fish are observed to be swimming actively and there is no noticeable abnormality in the behaviour, it means there is nothing to worry about the health of the fish in the aquarium. On the other hand, diseased fish may look off colour, restless and/or behave erratically. In such a situation, the fish should be closely observed for visible infection and remedial measures should be taken immediately. Stock improvement in terms of disease prevention must be a top priority issue. In order to achieve healthy fish species, it is necessary to have a health management protocols in every farming practices.

Common diseases of ornamental fishes

Common diseases of ornamental fish can be broadly grouped into four categories. They are:

1. Bacterial Diseases
2. Fungal Diseases
3. Protozoan Diseases
4. Parasitic Diseases

Table 7. List of common fish disease

Diseases	Causative Organisms	Clinical Sign and Symptoms
Cotton-wool disease	Fungus	White cotton woolly tufts on the body or fins
Eye disease	Fungus	Eyes become protruded from the socket & cloudy in colour.
Dropsy	Bacteria	Scale stand out the bloated body.
Gill rot disease	Fungus	Pale gills with signs of rotting, breathing rate much increased, loss of appetite.
Fin rot disease	Bacteria	Fins become split and rotten. The tissue between the fin rays gradually disintegrates
Melanosis disease	Multiple causes	Dark grey patches on the body which later turn black and peel living raw sports.
Lerniosis disease	Parasite	Threadlike work attached to fish
Tuberculosis disease	Bacteria	Fish hollow-eyed with a sunken chest. Sluggish air bubbles in fins. They appear to feed normally but gradually fade away until they are totally emaciated.
Shimmying Disease	Poor environment	Fish undulating from side to side without any forward motion are displaying typical symptoms of chilling. A rise in water temperature generally affects a successful recovery.
Ulcer disease	Poor environment	Open sores on the body. Loss of appetite.
Tumor Disease	Bacteria	This is a type of bacterial disease. Showing small pie-shaped tumour-like structure all over the body and gradually increase this towards large size.
Constipation Disease	Nutrition Disease	Lack of appetite and swollen belly.
White spot Disease	Protozoa	The appearance of small whitish spot or blister on the body. This disease has three cyclic life stages as a visible spot on the victim, like a dormant cyst and free-swimming period.
Argulosis	Parasite	The fish louse attaches itself to the host by suckers. Physical removal is possible but with reservation as for lernaea.
Cotton Mouth disease	Bacteria	The fluffy white growth around the mouth has given rise to the common name. If treated at an early stage the disease can be cured within a short time.

Bacterial diseases

Bacteria are responsible for many fatal diseases in fishes like furunculosis, columnaris, fin and tail rot, vibriosis, dropsy, cotton mouth disease and tuberculosis. The bacterial disease is the most common infectious problem of ornamental fishes. Collectively, only water quality problems exceed bacterial diseases in the area of pet fish morbidity and mortality. The majority of bacterial infections are caused by Gram-negative organisms such as *Aeromonas, Citrobacter, Edwardsiella, Flavobacterium (Flexibacter), Mycobacterium, Pseudomonas,* and *Vibrio*. The gram-positive bacteria also cause disease in ornamental fishes. eg. *Streptococcus*. Bacteria are either the primary cause of disease, or they may be secondary invaders, taking advantage of a breach in the fish integument and effects its immune system. The majority of bacterial fish pathogens are natural inhabitants of the aquatic environment, whether it is freshwater or marine. Almost all the bacterial pathogens of fish is capable of living independently away from the host. Any extrinsic stress, including shipping, crowding, poor water quality, and inadequate nutrition make ornamental fish prone to bacterial disease.

Bacteria are ubiquitous, present everywhere. A study of water from pet store aquariums showed wide ranges of Gram-positive, Gram-negative, aerobic, and anaerobic organisms. In fact, each of the eighteen samples of tropical fish water in this test contained *Pseudomonas* and *Citrobacter* while over 80% of the goldfish water samples contained *Pseudomonas, Citrobacter* and *Escherichia*.

Fin and Tail Rot

Fig. 114: Fin rot

Disintegrating fins that may be reduced to stumps, exposed fin rays, blood on the edges of fins, reddened areas at the base of fins, skin ulcers with grey or red margins, cloudy eyes. Possible predisposing factors: Poor water quality/ aquarium conditions and injury to the fin and tail. The affected area slowly breaks down. It is advisable to treat the water or fish with antibiotics @ 20-30 mg per litre. For mixing with feed, 1.0% of antibiotic can be used and fed to the fish. Antibiotics such as tetracycline is effective in controlling fin and tail rot conditions.

Scale Protrusion

Protruding scales without body bloat are seen. Scale protrusion is essentially a bacterial infection of the scales and/or body. An effective treatment is to add an antibiotic to the food. With flake food, about 1% of antibiotic such as chloremphenicol or tetracycline can be used. In the water, about 10 mg per litre of the antibiotic is required to be added.

Dropsy

Bloating of the body, protruding scales are found. Dropsy is caused by bacterial infection of the peritoneal area including kidneys, causing fluid accumulation. The fluids in the body build up and cause the fish to bloat up and the scales to protrude.

Fig. 115: Dropsy in rosy barb

Ulcerations, Red Sores or Eedpest

Bloody streaks on fins or body are found. Bacteria penetrate inside the body tissue and cause the disease.

Rearing water should be disinfected with suitable antiseptics such as acriflavine or onacrin (mono aminoacridine) with 0.2% solution @ 1 ml per litre followed by antibiotic treatment.

Table 8. Most Important Bacterial Diseases in Aquarium Fish.

Gram-negative bacteria	Gram-positive bacteria
Vibriosis (*V. anguillarum, V. harveyi clade, V. parahaemolyticus, Aliivibriosalmonicida (V. salmonicida), V. vulnificus , Photobacterium damselae)*	Mycobacteriosis *(Mycobacterium fortuitum, M. marinum, Nocardia asteroides, N. crassostreae (ostreae), N. seriolae)*
Aeromonasis (Motile Aeromonas spp.: *Aeromonas caviae, A. hydropila, A. sobria, A. veronii, A. jandaei; A. salmonicida)*	Streptococcosis *(Streptococcus agalactiae, S. iniae, Lactococcus garvieae, Aerococcusviridans)*
Edwardsiellosis*(Edwardsiellaanguillarum, E. ictaluri, E. piscicida, E. tarda, Yersinia ruckeri)*	Renibacteriosis *(Renibacteriumsalmoninarum)*
Pseudomonasis *(Pseudomonas anguilliseptica, P. fluorescens).*	Infection with Anaerobic Bacteria (*Clostridium botulinum, Enterobacterium catenabacterium)*
Flavobacteriosis *(Flavobacterium branchiophilum, F. columnare, F. psychrophilum, Tenacibaculummaritinum)*	Infection with Intracellular Bacteria *(Piscirickettsiasalmonis, Hepatobacterpenaei, Francisellanoatunensis, Chlamydia spp.)*

Fungal diseases

It is a common disease for hobbyists and pet shop keepers alike to refer to any grossly visible skin disease of tropical fish as "fungus", one frequently observed diseases such as lymphocystis and protozoal ectoparasitic diseases lumped into the fungus category. Fortunately, the fungal disease can be identified easliy under the microscope and other disease problems can be quickly ruled out, following a simple skin scraping. Fungi belonging to the genus *Saprolegnia* are the most commonly observed species affecting tropical fishes. Such fungi are opportunistic pathogens that typically colonize at exposed damaged tissue. A typical presence of *Saprolegnia* would be noticed on the fin rays of a catfish, which had been recklessly handled with a net.

It is noticed that most conditions if the fish is well supported with clean water and good food, the fungal tufts will be cured in time without any treatment. In severe cases, treatment is essential. These fungi are susceptible to several compounds including formaldehyde, malachite green and salt. Microscopically, one can see the typical fungal hyphae wound in a tight mat with the possibility of reproductive bodies being present.

Saprolegniasis

Saprolegniasis is a fungal disease of fish and fish eggs, caused by the *Saprolegnia* species called "water molds." They are common in fresh or brackish water. *Saprolegnia* can grow at wide range of temperature but prefers temperatures in the range of 15 to 30°C. It attacks an injured fish and from there it spreads to healthy tissues. Poor water quality with high organic loads is often associated with this infection. Saprolegniasis is often noticed by observing fluffy tufts of cotton-like material with a colour of white to grey and brown on the skin, fins, gills, or eyes of fish or on fish eggs.

Branchiomycosis

It is commonly called Gill Rot and is caused by the fung *Branchiomyces sanguinis* (carps) and *Branchiomyces demigrans* (Pike and Tench). Branchiomycosis is a pervasive problem in Europe, but has been only occasionally reported from other places as well. Both species of fungi are found in fish, suffering from an environmental stress, such as low pH, low dissolved oxygen, or a high algal bloom. *Branchiomyces* sp. grows at temperatures between 30 and 35°C. The main source of infection is the fungal spores carried in the water and detritus on pond bottoms. Fish may appear lethargic and may be seen gulping for air at the water surface. Gills appear striated or marbled with the pale areas, representing infected and dying tissue.

Ichthyophoniasis

Ichthyophonosis is a systemic granulomatous disease of freshwater and sea fish that is widely distributed geographically. The etiologic agent is

Ichthyophonus hoferi, a fungus-like protistan microorganism that has been taxonomically classified as a member of the class Mesomycetozoea, order Ichthyophonida (Mendoza *et al.*, 2002). Transmission appears to be direct and via the oral route, and infection is presumed to occur by the ingestion of contaminated feed, infected fish, and/or infected copepods (Roberts, 2012; Spanggaard *et al.,* 1994). The lesions of ichthyophonosis occur principally in highly vascularized organs, for example, the heart, spleen, kidney, and liver.

Parasitic diseases

Protozoan Diseases

The details of some important protozoan disease agents of tropical fish are discussed below:

Brooklynella

These are found in tropical saltwater and resembles *Chilodonella*. As they look very similar to *Chilodonella,* they are also known as cousins of *Chilodonella*. This parasite can be treated easily with protozoicides.

Chilodonella

It is a ciliated protozoan which can cause high morbidity and mortality among freshwater tropical fishes at the wholesale and fish farming levels of the industry. It attacks skin and gills. Easily identified microscopically by its heart-shaped structure and slow circular motion when not crawling on the surface of the fish. Once the disease is diagnosed, this problem is easily treated with formaldehyde, malachite green or salt solution.

Cryptocaryon

It is the marine form of "Ich." Frequently referred to as "white spot disease." The large ciliate possesses both free-swimming and encysted stages. Many aquarists take precaution to control this disease with the prophylactic use of copper sulfate in the water. Following the quarantine protocol for new fishes can greatly help to reduce the chances of spreading this disease.

Epistylis (Heteropolaria)

It is a freshwater stalked ciliate which is commonly found in water containing a high organic load. It tends to colonize in bottom-dwelling fish such as the plecostomus catfish. Lesions appear pale and white in colour and resemble a fungal disease. Microscopically, one observes a ciliated crown atop a long stalk which is prone to frequent contractions. It can be easily treated with formaldehyde. However, a clean well-filtered tank is the best solution to solve the problem. This disease is usually not fatal in itself but may open the fish up to secondary bacterial disease.

Henneguya

It is a sporozoan which is present in the form of small white cysts on the fins and gills of some fish. The cysts contain infective spores, commonly seen on the dorsal fins of imported *Leporinus* species. These are not harmful to the fish and can be carefully removed by scraping with a scalpel, which is the best treatment since the parasite is aesthetically undesirable.

Hexamita (Spironucleus)

These flagellated protozoa may cause severe gastrointestinal disease if present in large numbers. Normally, they inhabit in the fish digestive tract. As an ectoparasite, it is believed to be involved with "Hole in the Head Disease" (Head and Lateral Line Erosion) common to Oscars and other cichlids. It can be treated effectively with metronidazole.

Ichthyobodo

It is commonly called Costia, a flagellated protozoan ectoparasite. Microscopically they are very small (5-10 microns), move rapidly, and are sickle-shaped. They may be attached to host tissue or swim freely. Most common in freshwater species of fishes but has been reported from several marine fishes also. Generally inhabitant on fish skin. Poor water quality and overcrowding may allow this normally mutualistic parasite to reproduce rapidly and overwhelm the host. It can be treated with formaldehyde and malachite green but tougher than most protozoa.

Ichthyophthirius

It is commonly known as "Ich." The largest protozoan parasite of fish and one of the most commonly encountered species. Trophozoites may reach 1.0 mm in diameter. It can affect skin and gills or both. Prevention is the best method to control the parasite, although it is susceptible to a variety of parasiticides including malachite green and formaldehyde.

Plistophora

A microsporidian sporozoan and the causative agent for true "Neon Tetra Disease." The parasite is not specific to neon tetras and when present will attack the musculature of the affected fish. Infected muscle will contain numerous sporoblasts containing spores. The infected muscle appears white or pale. Certain bacterial skin diseases will produce similar gross lesions. Such sporozoan infections are usually unresponsive to treatment and diseased fish should be removed from the tank. High mortality is usually associated with this disease.

Tetrahymena

It is a ciliated protozoan which is free-living or parasitic in nature. It is a common cause of "Guppy Killer Disease." It is common in crowded conditions and in water containing excessive organic debris. These pear-shaped protozoa may be present in very large numbers when the infestation is severe. It is difficult to treat it by parasiticides because of its ability to burrow deeply into the skin of the host and get protection from chemotherapeutics. The best method to control it is prevention through best husbandry practices.

Trichodina

It is a disc-shaped ciliate protozoan found on the skin and gills of many fish. Circular rows of denticles and a ciliary girdle give this parasite a unique radial symmetry. It is not harmful when present in small numbers.

B. Trematode diseases

Both monogenean and digenean trematodes cause disease in tropical fishes. Monogenean parasites including *Dactylogyrus* and *Gyrodactylus* are ectoparasite and cause considerable damage to the host when present in big numbers. They possess a multiple hooked attachment organ called an opisthaptor which disrupts the integrity of the host's skin and mucus membranes.

These monogeneans can complete their entire life cycle on a single host. The life cycle may be as short as 60 hours if all environmental conditions are optimal. Crowding and other stress factors predispose tropical fish to monogenean trematode problems. These parasites are generally resistant to low doses of formaldehyde and even some organophosphates. Most freshwater monogeneans can be killed quickly with a 3 to 5-minute saltwater bath (30-35 parts per thousand). Glacial acetic acid or hydrogen peroxide dips will also kill these parasites. Praziquantel baths have also proved to be effective in killing some monogenean worms. While expensive, this is a relatively safe treatment when used at a concentration of 10 parts per million for 3 to 6 hours.

The majority of digenean fluke problems appear to be primarily aesthetic in nature among tropical fish. Fish commonly serve as an intermediate host for these parasites which frequently have a complex life cycle. Encysted digeneans are commonly observed as metacercaria in the skin and underlying tissues of tropical fish. Invertebrates may be the first host and a bird or mammal the primary host. Occasionally these metacercaria are found in the coelomic cavity of tropical fish.

Imported silver dollar fish species from South America are commonly infected with metacercaria belonging to the genus *Neascus*. Some fish may have only one or two metacercariae while others may harbour hundreds. This disease will not harm the fish and will not progress unless the fish is consumed by an appropriate primary host animal. Fish which are affected are sometimes said to have "Salt and Pepper" disease since the cysts become pigmented and the uplifted scales appear especially white or shiny. Another common digenean parasite is *Clinostomum* which is called the "Grub" by fish farmers in Florida.

Excysted worms may be more than 5 millimetres long and are easily visible to the naked eye. If the metacercaria are not too numerous, they can be removed safely with a clean scalpel. Occasionally, larvae belonging to the genus *Diplostomum* have been found associated with the lens in the eyes of tropical fish. In such cases the lens become opaque and the fish may be blinded. There is no reported treatment for this disease.

C. Cestode diseases

Tapeworms are found inhabiting the digestive tract of wild tropical fishes. Diagnosis can be made by fecal examination, observing proglottids exiting the vent of a fish, or during necropsy. Recently, work has been published, stating the use of praziquantel to treat infected fishes and it appears that certain tapeworms are susceptible to a dose as low as 2 parts per million in the water. Infected fish can be bathed in this solution for 3 hours with adequate aeration. Tropical fish commonly act as an intermediate host in a cestode's life cycle and encysted tapeworm larvae called procercoids can be found in the coelomic cavity of tropical fishes.

D. Nematode diseases

Nematodes are common parasites of fish and can be especially abundant in wild species. In some cases the tropical fish is the definitive host and the nematodes will be found in the gastrointestinal tract. In other instances the fish is an intermediate host and the larval nematodes will be seen encysted beneath the skin, in the musculature or in the coelomic cavity. Medical treatment of the larval forms is very difficult because these nematodes are encysted and well protected.

Some species of *Eustrongyloides* form large cysts just under the skin of tropical fish and can be removed surgically, especially if the fish is relatively large. As is the case with other encysted larval helminth parasites, the disease usually not progress unless the fish is eaten by the definitive host. Gastrointestinal nematodes can be observed on necropsy and ova are readily seen on examination of the faeces. While the presence of these parasites may not cause a problem in nature, the stresses of captivity and shipping may exacerbate any parasitic problem. Nematodicides such as fenbendazole and piperazine may be incorporated into food in order to successfully treat these problems.

E. Crustacean diseases

There are several important crustacean parasites of tropical fish. *Laernea*, commonly known as "Anchor worm," is a modified copepod parasite which infects large-scaled freshwater tropical and temperate species of fish. This parasite possesses a life cycle that includes microscopic pelagic larval stages that moult and grow several times before attacking the fish. On the host the female anchor worm matures and produces two large egg sacs, containing hundreds of *Laernea* eggs. This parasite is easily visible to the naked eye and may be more than 2 centimetres in length. They get their name from the attachment organ which is a highly modified structure which resembles the anchor on a ship. This structure found buried in the host's musculature and allows for the invasion of pathogenic bacteria.

Plucking the parasites from the fish is warranted and usually results in inflamed areas which heal quickly. Organophosphates and glacial acetic acid dips are successful in treating the problem. The disease is especially common in imported and domestic goldfish. The other major crustacean parasite is *Argulus*. This brachyuran crustacean is commonly called the "Fish Louse." Fish lice are flattened creatures with a very distinctive shape and appearance. They have a pair of eye spots and are about 5-10 millimetres in length. They move about the skin of a fish very effectively and camouflage themselves well on the host. They suck bodily fluids from the fish via a sharp stiletto that actually injects a small amount of toxin into the fish. These parasites are especially harmful to small fish. *Argulus* also possesses a life cycle with pelagic larval stages so the entire aquarium system may have to be treated with organophosphates to control the disease. Depending on the temperature, the total life cycle takes between 6 and 20 weeks.

A less commonly seen group of crustacean parasites are the isopods. While most isopods are free-living, members of the genus *Livoneca* can be parasitic. Terrestrial pill bugs commonly seen under rocks and logs are isopods and the aquatic parasitic forms resemble their land-dwelling relatives. While *Argulus* is dorsoventrally compressed, isopods like *Livoneca* are laterally compressed and appear segmented.

Viral diseases

Several viral diseases have been thoroughly described in ornamental tropical fishes. The most commonly observed viral disease of tropical fish is called lymphocytic disease. This disease is caused by an iridovirus, which infects connective tissue cells of the fish. The virus induces these cells to undergo extensive hypertrophy until the cells may actually be visible to the naked eye. Affected cells can increase a thousand folds in size. The disease appears to be more common in marine and brackish water fishes.

Certain species of freshwater tropical fish like the green terror (*Aequidens rivulatus*) are prone to the disease. Members of the genera *Scatophagus*, *Monodactylus*, and *Changa* are all brackishwater fishes that seems predisposed to the lymphocytic disease. Stress is almost certainly a factor in this disease since outbreaks are frequently observed, following capture and shipping of fishes. Gross lesions appear white and granular and usually are seen on the skin and fins. Occasionally, lesions are seen in the mouth and on the gills. There is no proven chemotherapeutic treatment.

Most cases are self-limiting if the fish is provided with proper water quality and nutrition. Surgery can be performed on affected fish by carefully scraping the hyperplastic fibroblasts to clear the fish with a sterile scalpel or scissors. This procedure should be performed quickly and the patient(s) should receive 5-10 days of topical antibiotic therapy, following the surgery. A definitive diagnosis can be made by microscopically examining a scraping of the affected area. The enlarged connective tissue cells appear circular and in clusters. These cells frequently emit a light orange hue under the microscope.

Ammonia poisoning

This is a common disease in a new aquarium when immediately stocked to full capacity. The amount of ammonia present in aquarium water is usually accompanied by an increased pH level. As ammonia is a strong base, it is stabilized by alkaline water. Ammonia can damage the gills at a level as small as 0.25 mg/L.

Gill becomes red; fish becomes darker in colour and grasp at the surface layer. Ammonia poisoning can be prevented but is impossible to cure. For water quality analysis ammonia and nitrite test kit test be used. Testing of the water should be done continously until the ammonia drops to nearly zero. At this time, one should notice an increase in the nitrite level. When the nitrites are not available, it will be safe to add fish. It is important to note that the bacterial phases not take place unless the tank is initially stocked with feeder fish which can be removed after treatment. For the immediate removal of ammonia, purchase of an ammonia detoxifier such as Kordon'sAmquel is advisable. However, it is best left alone until the bacterial load is sufficient.

Treatment of common fish diseases

The diagnoses method is perhaps the easiest method for a healthy ornamental fish. The fish must be removed from aquarium water while the parasite is removed. Periodical treatment is essential to prevent secondary infection caused by fungal or bacterial growth. Some of the common diseases are mentioned below along with the treatments. It is always safe to consult a fish pathologist before any treatments to fish.

Parasitic infection

Symptoms

The lice, worms, flukes are seen all over the body.

Treatment

Pluck the visible parasites from the fish. Follow with commercially available treatment such as aquatronic'sdiacide.

Fungal infection

Symptoms

Appearance of white colored cotton-like substance concentrated mainly on scrapes, Surface injuries, fins or mouth.

Fig. 116: Saprolegnia fungal disease

Treatment

The treatment for this disease is relatively easy. There are a great many commercially available products for this, including maroxy and super sulfo and would be controlled by aquatronics.

External bacterial infection

Symptoms

Many red or orange spots appear on the body. Dropsy is also a sign of a bacterial disorder. The false fungal infection looks like fungus but it is originally a bacterial disease known as columnaris. This symptom may include a grey or white film-like layer on all over the body.

Treatment

There are a number of effective treatments for many bacterial infections. The main treatment can be made by tetracycline, penicillin and nalidixic acid. NaCl solution can also be used for this infection as a bath or dip treatment.

False Neon Disease

Symptoms

Pale white patches visible under the skin beneath the dorsal fin. The fish also becomes sunken. The fishes show erratic swimming movement.

Treatment

Some antibacterial remedies are successful; correction of environmental factors prevent re-infestation.

Swim bladder disease

Symptoms

Affected fishes generally unable to maintain their body balance. The fish swim uncomfortably. Additionally the fish may sink to the bottom of aquarium and rise with great effort.

Treatment

In case of this disease prevention method is better than a cure. A mild salt bath or feeding of live foods can prevent this problem.

Velvet disease

Symptoms

Affected fishes rub their body against hard objects. A yellowish-grey film like layer with tiny spots appears on fish's body. The fishes show respiratory problem and they lose their appetite.

Treatment

Application of commercial chemical treatment containing copper is most effective. Dimming the aquarium lighting and subjecting the effected fish with a strong salt bath (3% NaCl solution) also can recover this infection. It is worth treating up to one week after all signs of velvet are removed from the aquarium to ensure that mature parasite have been discharged from the fish.

Hole in the head

Symptoms

Initially small holes appear on the head with a tiny white parasite protruding in the advance stage holes become larger as the skin is eaten away. The fish may also show more intense colour. Fish start swimming erratically and show sign of malnutrition.

Treatment

Specific anti-parasitic medication is available to treat this disease which is also known as Hexamita. Metronidazole medicine (trade name Flagyl) with food to tackle the parasite can also be used.

Black spot disease

Symptoms

Small black spots are visible clearly on the body.

Treatment

This disease is generally easy to cure. The infected fish should be given a freshwater dip, followed by a formalin bath and continue treatment until recovery. Methylene Blue has been used with some success in quarantine tanks, but not in the main tank as it will damage the biological filter. There are a number of commercially available treatments and preventatives are available in the market.

Cataracts

Symptoms

White or grey colored layer like structure covering the eyes only.

Treatment

The only treatment specifically designed for this disease is removing eye fungus by aquatronics. In the process, special attention should be made to assure that ammonia and nitrite levels stay within accepted measures.

Cauliflower disease

Symptoms

Small white spot appear on the body and the fins. As the name suggests, they develop into the small cauliflower-like structure and eventually, infected fish deteriorate and finally die. However, it is observed in some cases that infected fish recover quickly without any treatment.

Treatment

Treatment of most viruses is uncertain. It is suggested that trimming the affected portion of the fins may help followed by several bath in solution of Malachite green. Disposal of the affected fish is the only other option.

Diagramatic Representation of Aetiology of Disease

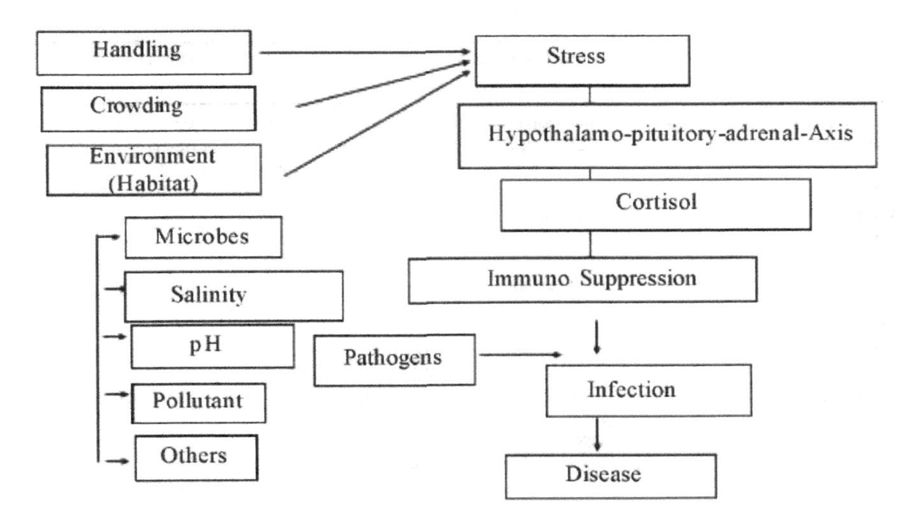

Approved aquaculture drugs

1. Formalin solution
   - Paracide-FR
   - FinquelR and Tricaine-S
   - Parasite-SR, Formacide-BR and Formalin-FR
2. Florfenicol
   - AquaflorR Type A Medicated Article
   - AquaflorR CA1
   - Tricaine methanesulfonate (MS-222)
3. Oxytetracycline
4. Terramycin R 200 for Fish (oxytetracycline dihydrate) Type A Medicated Article

5. Oxy Marine™, Oxytetracycline
6. HCl Soluble Powder-343,
7. Terramycin-343, TETROXY Aquatic
8. Hydrogen peroxide 35% PEROX-AIDR
9. Sulfamerazine
   - Sulfadimethoxine/ormetoprim combination
10. Romet-30R

Table 9. Common fish diseases and treatments.

S. No.	Symptoms	Disease	Causative agent	Treatment	Observations
1.	Pin head size white spots on the body and fins	Ichthyoph-thyriasis	Protozoan parasites (Ichthyoph-thyrius)	Increase the temperature and treat with 5% Methylene blue (1 drop/L). Treat the fish for 1 week	Contagious observed during sudden drop of water temperature
2.	White spots smaller than above	Oodinium	Unicellular parasite (mono flagellate)	Copper sulphate 0.1 mg/L for 10-15 minutes bath	Contagious
3.	White clumps with cotton like appearance	Saprolegniasis	Fungi including *Saprolegnia*	1 tsp salt/ 2 L water, 1-2 drops of Methylene blue or Malachite green 2 mg/L for 30 min. dip or 0.1 mg/L for permanent bath	Favoured by wounds
4.	Swollen eyes	Exophthalmus	Bacteria, viruses, fungi, sometimes together	1% Silver nitrate on poped eyes followed by 1% Potassium permanganate	Treatment is difficult
5.	Gradual disintegration of fins	Pseudomoniasis	Bacterial disease	Surgical removal of ragged portion by a fine sterilized scissor, paint the cut wounded portion by Iodine solution. Repeat the same at 12 and 24 hours	Unusual swimming behaviour
6.	Swollen abdomen, erected scale	Dropsy	Bacterial disease	No known cure. Antibiotics may be tried	Contagious, difficult to treat
7.	Opercula sticking out, usual swimming, tiny flukes on the gills	Gyrodactyliasis	Parasite fluke, Gyrodactylus	Formaldehyde 5-6 drops/L water dip treatment for 10 minutes. Repeat for 3 days	Not east to detect
8.	Red patch in body,	Argulosis/ learnaeasis	Argulus/ Lernae Ectoparasite	Physical removal of parasite, 15 minutes bath in 1-2% potassium permanganate, painting the region with iodine solution.Repeat the same after 12 and 24 hours	Violent rubbing due to irritation
9.	Unusual swimming behaviour	Longer exposure to poor water quality	Oxygen problems, poor water quality due to nitrogenous substances	Increase of oxygen, one third water exchange, check pH and nitrates level	Symptoms can also correspond to an infectious disease.
10.	Gasping for air at the surface	Lack of oxygen	Defective aeration,	Increase the aeration, check all water parameters	This can be a symptom of infectious disease.

11.	Weight loss and poor growth	Nutritional problems	Underfeeding or lack of balanced diet	Fresh feed and live food	Possibility of commercial vitamins in the feed
12.	Skeletal deformity	Hereditary problem or lack of vitamin C in the feed	Genetic origin of Nutritional disorder	If mass deformity occurs separate the parents, check the Vitamin C level.	It is not unusual for a few of the fry in a batch to be affected

Table 10. Adverse environment conditions and related problems.

S. No.	Problem	Symptom in fish	Remedial measures
1.	Chlorinated water	Restless movement, loss of balance	Vigorous aeration for 24 hours.
2.	Ammonia load	Inflamed gills, and fin edges blood spot, loss of balance	Water exchange and aeration required.
3.	Nitrites and Nitrates	-Do-	Partial water exchange
4.	Oxygen and nitrogen (excess)	Gas bubble disease. Small bubbles visible under skin, in fins and around eyes, Exophthalmia (Pop eye)	Keep the tank away from sunlight, if algae or plants exist.
5.	Insufficient oxygen	Gasping at surface, loss of body color.	Vigorous aeration is required with partial water exchange.
6.	pH		
	(a) Acidosis (pH 04-05)	Fast swimming movements, gasping at surface. Fish jump out from water	Avoid over stocking, Partial water exchange. Stop feeding temporarily.
	(b) Extreme Acidosis (pH>04)	Sluggishness tendency to settle in bottom and to hide, loss of appetite and colour.	Stop feeding, exchange water, and provide aeration.
	(c) Alkalosis pH(09-10)	Serious damage to gills, disintegration of fin edges, body slime is excessive.	Avoid direct sunlight and partial water exchange.
	(d) Temperature (low)	Sluggish movement, resting at bottom. Fin and body movement gradually allows down. Loss of coloration.	Add ground water drawn through deep bore well, as part of water exchange exercise.
	(e) Temperature (high)	Increased level of activity (above normal) Increase in metabolism, increased rate of respiration, gasping at surface.	Addition of water or in an extreme situation thermostat also can be used if the tank is small. Add well water. Water exchange is very much essential.

Tips

1. To avoid stress during transportation Valium @ 5-6 mg/l can be used. By applying this drug fish may survive for long time.
2. In case of long-term transportation one Aspirin tablet in 15 liter transporting water can be applied. For this application stress condition should avoided during transportation.
3. To avoid stress during transportation Valium @ 5-6 mg/l can be used. By applying this drug fish may survive for long time.

4. Sometimes aquarium fishes are affected from fungal disease. The body of the fish gets covered by fungi. We should apply Ceaetoconazol or Fluchonazol group's medicines @ 1 tablet in 15 l of water for bath treatment (15-20 minute).

11

Handling, Packaging and Transportation

Ornamental fish breeding and trade provide excellent opportunities for employment and income generation. One of the most critical determinants of the success of the trade is the problem of delivering a quality product to the market. The know-how of collection, critical handling, packaging methods are essential for serving the successful transport of live tropical fish to their final destination. However, a major impediment in expanding the trade of indigenous fishes is the poor roads/communication from most of the collection/landing centres to the town, non-availability of oxygen packaging at remote collection centres, difficulties in transportation of the collected fishes to the agent/stockist. Transportation of live fish and shellfish from areas of collection to destination or from areas of farming to the destination is an important activity of ornamental fish industry.

Proper collection, packing and transportation may help in increasing trade to a greater extent. Ornamental fishes destined for marketing, both domestic and export, have to use modern post-harvest technology or modifications of the current post-harvest technology to improve the post-harvest and post-shipment survival that is critical to the industry. It is in this context that the significance of packing and transport come into the picture.To overcome the problem of high mortality of fish species at different stages during transportation, it becomes essential to evaluate the most suitable and ideal condition for transportation. At present, the mortality rate during fish catching, collection, transportation is very high. The claim by the importers due to DOA (Death on Arrival) from India is on the higher side compared to the consignments from other developing countries. This is due to the lack of appropriate technology for fish packing & transportation/shipment. Similarly, since the airfreight charges are very high, the exporters have no idea of the optimum number of a particular live fish to be packed in a container to reduce transportation cost. Such information is lacking for Indian fishes. The success of ornamental fish transportation depends on the following steps:

Collection of ornamental fish

This is the first step in the trade and is important to overcome the problem of high mortality of fish species at different stages of transportation. At present the mortality rate during fish catching, collection and transportation is very high, which is due to lack of proper know-how. Therefore careful handling of fishes right from its collection is very essential. Generally, various traps and gears viz., shore seining, (ghera Jal) electrofishing, dipnets, encircling net and brush fishing are practised in the river, floodplain wetlands and low-lying water areas of the eastern and northeastern region of India. In addition, numerous indigenous gear are also used in marginal areas of rivers, wetlands etc, to collect the fishes alive. In the collection of wild fishes, one must be aware of the water body and site from where the maximum number of fishes could be collected. While collecting the fish the following points must be taken into account:

- Selection of species having actual value/potential in the export market.
- Screening of potential species for different traits like the ability to withstand handling/transportation stress etc.
- Trial to keep them alive both on collection site and in stocking site.
- Care should be taken in the collection and handling of fishes with minimum stress.
- The gear and fishing method should be selected as per the habit of the fish. Generally, soft small mesh nylon nets or traps are used for catching the small fish. After collection, the fishes are kept in nylon hapas fixed in the same water bodies for 1-2 hours. Then, they are brought to the shore and kept in a container provided with aeration.

Acclimatization /conditioning

It is a process by which fishes are accustomed to live in an artificial condition with crowding in limited space and oxygen. It is very much important for taking up any fish for the transportation programme. By this practice, not only the fishes are trained to live in a crowded condition but also release a less excretory substance in the transporting medium. Wild collected fishes need to be acclimatized for about 10-15 days. This is necessary to condition the fish in a changing environment (pH, hardness, temperature *etc.*) It is necessary to record the water temperature of the collection site and maintain the same in the conditioning water as far as possible. Conditioned water should be better if it is aged for two to three days with tetracycline treatment; this is a must for tap, reservoir, and pond water. It can also be done by methyline blue treatment with common salt and Epsom salt, which helps to bring out the slime on the fish and prevent the fish from damage/stress. Acriflavine treatment is found to be more effective in checking the bacterial and fungal disease outbreak during transportation. After initial quarantine and conditioning the fishes are placed in a holding tank having clean, fresh and well-aerated water.

Sorting of fish

Though it is labour-intensive but necessary to ensure the quality of fish to be marketed. The harvested or collected fish that have already been sorted by size are collected from holding/stocking tank are anaesthetized. Once properly sedated they are sorted by sex and colour and placed in different tanks till packing. Small scale sorting can be done without the sedative/tranquilizer.

Methods of packaging

Mainly there are two ways of fish packing.

a) Open system

An open system, comprising open carrier with or without aeration, is used for transporting fish at a short distance. Earthen or aluminium hundies are used for carrying the fishes. Depending on the size, 50,000 to 75,000 fishes can be transported at a time in 23-32 l hundies. To increase the dissolved oxygen content of the water, the traders hit the surface water of the hundies containing fish continuously by their hand. Sometimes they also exchange part of water with fresh and clean water, depending upon the condition of fish, temperature, stocking density and distance to be travelled. The mortality percentage may vary from 5-20. It is better to transport the fishes in cool hours i.e. either in the early morning or in the evening.

b) Closed system

Closed system is used for transporting the fish to long distances in a sealed airtight carrier with oxygen. The fishes are carried in polyethylene bags filled with 1/3 of clean water and 2/3 oxygen.

Packaging films

The packaging films that are commonly availabe in the market include LDPE, LLDPE, HDPE, HMHDPE, PCC, NYLON-6, PS, PC, etc. Low-density polyethylene (LDPE) is the most commonly used material since it possesses many desirable qualities such as transparency, water vapour impermeability, heat sealability, chemical innerness and is fairly economical. But the organic vapours and oxygen and carbon dioxide perme abilities are high and LDPE has poor grease barrier property. It resists temperature between 40°c to 85°c. Linear low-density polyethylene (LLDPE) has better mechanical strength and heat sealability than LDPE. These properties can be made use of by blending with LDPE.

Factors to be considered during packaging

a) Density

Under carefully controlled conditions densities up to 1 kg of fish per litre of water have been successfully transported. Densities of around one-tenth of this are more usual and ensure that the health and well being of the fish are preserved. As an appropriate guide 2 kg of fish can be placed in 20 litres of water inside a polythene bag, with large oxygen-filled space above it, and at 10 degree Celsius can be carried for 5h without the need for further oxygenation. However, the density chosen will in practice depend upon the species, the type of the tank, the temperature and many other factors, and must really be determined by trial and error for each specific situation. At high densities fish become agitated which increases oxygen consumption and risk of damage. Loss of mucous may also occur which can cause discomfort or choking of the gills. Foaming of mucus laden water may occur with aeration, and may be suppressed by the non-toxic antifoaming agent.

b) Temperature

Temperature influences the physiological activity and the oxygen consumption of the fish, and also the oxygen-carrying capacity of the water. High temperature may also be directly lethal to fish. From all these aspects a low water temperature at least as cool as the water from which the fish are taken is preferred. Cooling the fish has often been used successfully to calm fish for transport. Chipped or crushed ice is satisfactory for most of the journeys but not for long-distance air freighting of fish, dry ice has a greater cooling capacity for its weight. However, caution must be exercised to prevent from the evaporating CO_2 from dry ice coming into contact with the water. Deep frozen blocks of ice or special cooling bags are safer and last for a very long time.

When fish from warm tropical waters are taken for packing, water-cooling can be a handicap. Heavy insulation reduces cooling to a minimum and chemical-heating packs can be helpful. The problem is most pronounced when tropical species are being transported to temperate climates, especially if fish are transshipped from one aircraft to the other and the container is left exposed to ambient conditions at the airport. Perhaps the best solution in this event is to arrange someone on the spot to properly look after fish being transshipped.

In most cases, extreme temperature changes can be avoided by adequately insulating the fish container with plastic foam or expanded polystyrene and by transporting only at the appropriate time of the day. Temperature changes should be effected gradually. For example by switching on to cool gear or adding ice bags or dry ice containers after the fish have been placed in the transporting tank. On arrival at more permanent holding facilities, the temperature should be raised or lowered slowly over several hours if the difference is more than a few degrees.

c) Dissolved gases

To maintain fish in healthy state, there must be sufficient oxygen in the water. In addition, the build-up of carbon dioxide and ammonia must be prevented. Gas concentration can become critical under transport conditions where the fish are stressed and their oxygen requirement is greatly elevated. Oxygen concentration can be maintained by bubbling compressed or pumped oxygen or air, or by surface agitation. For small quantities of transport for a short period, continuous aeration is necessary. The fish are provided with well-aerated water with an air space above it. The vehicle movement will provide an agitation effect. The effectiveness is increased by maximizing the surface area, or by providing an atmosphere of oxygen above water. For this purpose, a large space of about 4 times water volume must be left in the container to contain the air or oxygen. Where larger fish are being transported on especially long journeys or when animals have a large oxygen demand, it is necessary to bubble air or oxygen continuously through the tank. Pure oxygen is not harmful to freshwater fish but the cost of gas and associated equipment may render compressed or pumped air a more attractive alternative. If a gas cylinder is used, it is important to use a pressure regulator to meter the gas flow and to ensure that gas continues to flow at a constant rate regardless of the cylinder pressure.

Carbon dioxide is toxic to fish, both directly and by decreasing their ability to extract oxygen from the water. With low densities and high aeration rates, it is unlikely to attain toxic levels. Where fish are transported without continual aeration, CO_2 does build up. With high densities of fish and low densities of oxygen, CO_2 concentration may become a problem though the level of oxygen remain high. Loss of balance of fish can occur at CO_2 level below lethal, so advance warning of a critical situation is given.

Ammonia is produced as the major nitrogenous excretory produce by most teleosts and is very toxic. Oxygen concentrations and pH both affect ammonia toxicity. It is the unionized ammonia which is particularly toxic and the equilibrium is markedly influenced by pH. A shift from pH 8 to 7 produces a ten fold decrease in the quantity of unionized ammonia. The decrease in dissolved oxygen increases the toxicity of un-ionized ammonia. A change in pH ratio across a tissue (eg gills) can also greatly influence the concentration of the un-ionized gas on each side of the barrier. At low fish densities the build-up of ammonia is unlikely to create problems during transport. The risk can be reduced further by holding fish without feeding for two days before transporting as ammonia excretions rate drops rapidly in the case of unfed fish.

d) Salinity

The body fluids of saltwater and freshwater fishes have a salt concentration between those of freshwater and seawater. Thus, both are under osmotic stress and are having to work to maintain their internal ionic equilibrium. When fish is physically damaged, the rate of exchange can increase and represent greater stress. Alteration of the external medium has been used in the transport of both freshwater and sea water fish to reduce this stress.

e) Anesthetics

Increased physical activity during transport can adversely affect the health of the fish in two ways. First is physical damage by the abrasion with the packing container; second is by a physiological reaction to physical activity and other environmental factors such as low dissolved oxygen. Such reaction is manifested in high blood lactate levels, which can cause serious debilitation or death. The level of physical activity of the transported fish must be kept to the minimum. Lower temperature can help, as can covering carrying tanks with light proof material. A third possibility is the use of tranquillizers. A considerable range of chemicals are potentially useful as tranquillizers, some being employed at higher concentrations as anesthetics.

Several anesthetics now find application in the transportation of fish seeds. Though the concentration of various anesthetics, used, differs widely depending upon the species, size, etc. they have common characteristic feature, viz., lowering the metabolic activity of the fishes by their depressing action on the brain. This actually leads to the low consumption of oxygen in the media during transit and thus facilitates higher survival and long duration of journey in oxygen packed containers under higher stocking densities.

The sedating of fish brings in practical benefits by way of:

a) Decreasing the rate of oxygen consumption and reducing the rate of excretion of carbon dioxide, ammonia and other toxic wastes.

b) Controlling the excitability of the fish and thereby reducing chances of injury.

c) Reducing the time required for handling them.

Any anaesthetic that is used for the transportation of fish must possess the following:

- The drug must be water-soluble
- Dosage required should be low
- Time of induction and recovery should be short
- Fish should tolerate well for several hours at low concentrations
- Should not have any side effects on the fish
- Lethal concentration should not be high, so that, fish do not die accidentally.

The most inexpensive method of tranquilising fish is the use of water chilled down to 5-10 °C without the use of any drugs. The common tranquilisers, used in live fish transportation, are Amobarbital sodium, Barbitol sodium, Chloral hydrate, Hydroxyquinaldine, Tricaine methanesulphonate: MS 222, Methyl paraphynol (Dormison), Quinaldine, Sodium amital, Tertiary amyl alcohol, Thiorucid, Urathan, 2-phenoxy-ethanol, etc. The dosage of these tranquillizers has to be standardized for each species and packing density to avoid any adverse effect.

Packing process

Fishes to be packed are sorted a few days earlier and kept in clean water to adjust them in the transporting media. They should not be given any food before packing. Desired number of fish is placed in the plastic bag, depending upon the size. The upper part of the plastic bag is then compressed to drive out the air and inflated with pure oxygen under pressure. After filling the gas, left out portion of the plastic bag is twisted and folded and tied tightly by rubber band or thread. The plastic bags, containing fish, are now ready for packing in insulated cardboard or styrofoam boxes for transportation in normal cargo by road, rail or air. The fishes can be transported safely and with minimum mortality up to a period of 72 hours. What all needed in this process are: plastic bags, (the quality of plastic should be high oxygen retentive and bear tensile and tearing strength), insulated boxes, pure oxygen and rubber band or thread.

Pack number is based on the number of pieces per 7 lit of water in a single bag. The packing density depends on the size of the fish. Normally, egg layers are packed with a 5% overpack and livebearer with 10% overpack. For eg., A box of mollies shipped at 300 pieces actually contains 330 pieces. Higher priced items, extremely delicate and large fish such as cichlids are not over packed. Large fish with sharp spines or scales should be placed in double bags to reduce the possibility of puncture and leakage.

Causes of fish mortality

The mortality of fishes during transportation may be caused due to unplanned management made before transportation such as unstarved fish, high-density packing, the inclusion of unhealthy/diseased fishes etc. The followings can be other causes of mortality during transportation:

1. High Carbon dioxide accumulation or deficiency of oxygen.
2. Ammonia or another accumulated metabolite in the medium.
3. Improper handling of fish while collecting, sorting and packing and after receiving.
4. Quick transfer into a new medium without proper conditioning.
5. Injury and predation.

Use of medicines/chemicals in transportation

Tranquillizers and disinfectants are used in transportation to prevent mortality of fishes. The sedatives bring down the oxygen consumption rate and reduce the rate of carbon dioxide excretion. It also controls the injury and predation rate of fishes.

The commonly used anaesthetics are sodium amytal (50 ppm), Tertiary amyl alcohol (2 ml/gallon of water), Dormison (1-2 ml/gallon), MS 222 (50 ppm), Urethan (100 ppm), Triouracil (10 ppm), Quinaldine (5-10 ppm) and Hydroxy quinaldine (1 ppm)

The antibiotics are used for prophylactic and quarantine control of fishes. A bath before packing of fish or addition in the container water is recommended. Acriflavine, methylene blue, copper sulphate, potassium permanganate, chloromycetin and formalin are used for the purpose.

Preparation of export consignment

To ensure maximum survival of fish during transportation, various management tools are employed such as:

- The fishes should be starved for 24-48 hrs to empty the digestive tract. This reduces ammonia production and faecal excretion during the transportation.
- The water temperature should be reduced to slow down metabolic activity, thereby reducing oxygen consumption and ammonia production.
- Fish are packed following IATA (International Act of Transport Authority) regulation, which states that fishes must be able to survive up to 48 hours in the container.

- Fish are packed using a double bag system where one polyethylene bag is placed inside the other. The inner bag is filled with 1/3 of water at 10 °C and 2/3 pure oxygen. Both bags are sealed separately with rubber bands and then placed inside the insulated cardboard or poly boxes.
- Ice or dry ice (Solidified carbon dioxide) may be packed around the outer side of the transport bag, containing the fish to help to maintain the low water temperature, particularly in warm weather.
- Zeolite can be used to absorb the ammonia produced by fish during transportation.
- Water pH often drops to 6 or less due to production of CO_2 in the process of respiration, which can be minimized by the use of suitable buffer.
- Transport to the airport should be in temperature-controlled vehicles. At the airport the boxes should be kept out of direct sunlight to prevent temperature increases.
- Strict regulations dictate how the boxes should be handled and placed in flight. Generally, direct flight are preferred to avoid delays in the change of subsequent flight.
- Appropriate documents must accompany the consignment.

Fig. 117: Packaging area

Fig. 118: Oxygen packing of ornamental fish

Fig. 119: Oxygen packed fish ready for sale

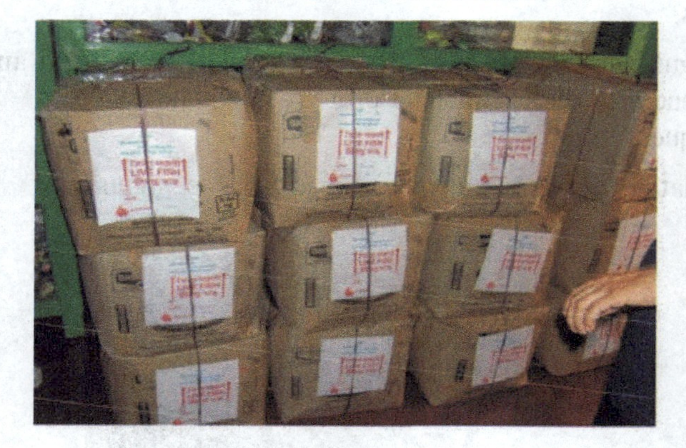

Fig. 120: Cartoon packing for long distance delivery

Package design for exports

Package designing is a unique technological philosophy meeting the various requirements in the overall distribution chain of the manufactured or processed item. Within this definition, several functions can be identified - protection, and preservation during transit, handling and storage, marketing strategies, rules and regulations, etc. Thus the importance of packaging must be reckoned in the early stages of production and marketing. Product and packaging are inseparable.

Packaging materials are selected to suit the nature of protection needed, its anticipated shelf life, and their compatibility with the product. Shipment packages are the external packages containing the unit containers. These transit packages offer protection against handling and transit hazard. The following points need consideration in designing the transport container:

1. Protection - Against transport conditions
2. Handling - Rationalizing of the package size for handling in transit and storage
3. Shipping - Conditions of shipping rules, regulations, freights, tariffs, journey hazards
4. Manufacturing - To suit in-plant packaging line
5. Functional utility - Convenience for transportation staff, handling, labour, carrier space.
6. Identification - International markings on the package
7. Disposability - To avoid public hazards and pollution
8. Economy- Packaging costs

The strength and quality of plastic bags, single or double use of plastic bags, the type of insulation lining if corrugated cardboard box is used, the quality of all-foam boxes, the quality and quantity of the water in the water bags, the number and type of fish placed in the single bag, the amount of O_2 in each bag and last but not the least the preparation of the consignment before shipment are all the factors that affect the safety and survival of the fish during transport and transit.

Transportation

In recent years, transportation of live fish and shellfish is becoming an important activity of the fish industry. With the development of aquaculture, transport of cultured live fish fry, fingerlings and post larvae by road, water and air from hatcheries to nearby farms and other states and countries is increasing. Trade-in the export of ornamental aquarium fish is a multimillion-dollar business. The success for all of which depends on effective packaging techniques and careful handling practices prior to and during shipment with minimum mortality.

Transportation of live fish and shellfish from areas of collection to destination or from the area of farming to the destination is an important activity of the ornamental fish industry, and the success of transportation depends on:

- Careful handling practices right from harvesting /quality assurance,
- Conditioning for a period of time before packing,
- Effective packaging techniques,
- Stocking density,
- Maintenance of low temperature and high humidity during storage and transit,
- Minimum mortality,

- Replenishment of air with oxygen and reduction in the accumulation of toxic wastes and controlling acidity of water medium with suitable buffers,
- To reduce the metabolic activity and thus their oxygen consumption,
- Conditioning the fish to the environment prior to transportation/shipment,
- The lower temperature of the water, and
- Addition of permissible and correct dosage of anesthetics.

If not properly planned and implemented, large scale mortality would occur resulting in heavy loss. The most important reasons for the mortality of fishes during transportation are:

- Mistakes made before transportation (Fed fish, fish density too high, the inclusion of weak and diseased fish, etc.).
- High carbon dioxide tension and or/deficiency of oxygen in the transporting medium.
- Toxicity of accumulating wastes like ammonia and other metabolites in the medium.
- Improper handling of fish while collection, packing and after arrival leading to physical injuries.
- Too quick transfers into new water without proper acclimatization, wrong treatment for diseases affecting the health of fish.

The duration of transportation and temperature during transportation and its fluctuation are the two most critical factors affecting the survival of ornamental fishes. Standardization of the handling and transportation procedures with respect to temperature and time is hence essential to develop a successful export industry for ornamental fishes. Similarly, the handling of the fish during catching, conditioning, packing, transportation, etc. are also crucial. The stress and activity of the fish should be kept at a minimum.

12

Biosafety and Hygiene

The purpose of managing a fishery is to maintain the resource so that it is renewable and therefore sustainable. Responsible aquaculture practices rely on sustainable production systems, to minimize impacts on the natural environment, and support resource conservation. In other words, the harvest of fish from the wild or their domestic culture, if performed with sound foundations of ecological and economic principles, can be sustainable and self-reliant commercial industry. In traditional subsistence fisheries, fishermen use primitive and inefficient gear to capture most ornamental fish. However, the supply of aquarium fish is not inexhaustible, and signs of over-fishing are becoming apparent in localized areas. With high demand and pricing of many beautiful species, ornamental fish are being harvested at greater volumes and higher rates, threatening the viability or sustainability of the fishery. Most endemic and sensitive species are restricted to very narrow specific habitats. Their survival, affected by physical over-exploitation for the aquarium trade, may be further hit by habitat alteration. No comprehensive studies have been carried out on the requirements of these endemic or sensitive exported species. In the absence of suitable impact studies, it is not possible to predict the impact of habitat alteration on these species. Exporters target more colorful varieties and since their ecological significance has not been studied, what long-term effect such selective exploitation will have on genetic diversity cannot yet be defined.

Apart from legislation that can be effectively implemented, eco-physiological and population studies of a quantitative nature are urgently needed to advise on the collection, maintenance and transport conditions that need to be followed by exporters to safeguard collected stocks from unnecessary mortality. Exporters are eager to learn and would be receptive to receiving appropriate, scientifically formulated, well-meaning practical advice. Studies should be targeted towards this end since it seems unlikely that the export trade can at present be voluntarily modulated on the basis of conservation requirements. Such a strategy could only become feasible after an adequately robust ecological database has been compiled, which would necessarily require time.

An effective management strategy needs to address not only aquarium-trade related matters but also policy and other matters in an integrated approach if we are to be hopeful of sustaining the ornamental fish industry in the long-term.

Regulations in ornamental fish trade in India

The Ministry of Environment, Forest and Climate Change has banned the display and sale of 158 species and has made mandatory the appointment of a full-time fisheries expert for monitoring the health of the fishes in the tank, apart from bringing out rules on tank size, volume of water and stocking density. The following regulations have been recently set up through government initiatives in the country for aquarium and fish tank animals shop:

i) No aquarium or fish shop shall source fish tank animals, caught by destructive fishing practices such as bottom trawling, cyanide fishing, use of explosives or dynamite to kill or stun fish, those trapped from coral reefs or from any protected areas.

ii) The rules require every aquarium to acquire recognition and fish shop to acquire registration from Animal Welfare Board of India.

iii) Rules prohibit aquariums & fish shops from keeping, housing, displaying and trade of cetaceans, penguins, otters, manatees or sea or marine turtles, artificially coloured fish, species protected under Wildlife Protection Act, 1972, species listed under Appendix I in CITES and over 150 species listed in Scheduled II of the rules.

iv) No aquariums will be allowed in public areas such as airports, public offices, railway stations or schools nor in markets or exhibitions where permanent facilities are not available.

v) Prerequisite for recognition is submitting application to the state board along with the blue print and collection plan. Certification for obtaining recognition an aquarium is required to submit a master plan (blue print and collection plan) to the State Animal Welfare Board. The recognition from AWBI is subject to approval of the state board as well as itself. Recognition to be renewed every 2 years.

vi) The rules prescribe, for aquariums, standards for a veterinary and infrastructure facilities. They mandate maintenance of inventory (record of births, acquisitions, deaths), feed register and health register; fish shops in addition to aforementioned records shall maintain an inventory of animals traded (sold, bought exchanged) during the previous year.

vii) Rules prohibit physical handling or performances by fish tank animals as educational activity.

viii) Noncompliance of the rules by aquarium will result in de-recognition and consequently sealing of aquarium by the local authority. Animals confiscated from such aquarium are required to be handed over to a recognized aquarium or be released into their natural habitat. Noncompliance of the rules or failure to apply for the registration within stipulated time will also result in the sealing of the fish shop by the local authority upon recommendation of the State Board.

ix) Certain common provisions for aquariums and fish shops with regard to housing and display prohibits display of fish tank animals in bowls or in tanks of the capacity of 60 litres or less. These provisions prescribe conditions for maintenance of fish tanks, enrichment to be provided in fish tanks and standards for upkeep and health care of fish tank animals.

Impacts of exotic ornamental fishes on aquatic ecosystem

Several studies have clearly emphasized that alien fishes frequently alter the aquatic ecology by changing water quality (e.g. increase in nitrogen and phosphorus concentration) and also cause the extinction of native fishes by predation (destroying the eggs, larvae, sub-adults and adults), damaging the aquatic vegetation and exploiting the food resources. Besides, a number of alien fish species also hybridize with indigenous species in the wild, diluting the wild genetic stock leading to long-term introgression of gene pools. Mostly, the invasion of aquarium fishes triggers the native species decline and ecological destruction of the native system. However, in India, there is no detailed study which discusses the impacts of ornamental fishes at every trophic level. Majority of the Indian studies reported about the occurrence of the species in the inland water rather than its detailed impacts on the system. Besides, it is worth mentioning here that the decline of native fish variety will affect the livelihood, health and general well being of the rural and indigenous community. In order to create awareness among different groups about the deleterious role of extotic aquarium fishes in the wild, this article discusses some of the important impacts (biological/ecological) of ornamental species reported in several parts of the world and in India as well.

Wildlife Act in relation to ornamental fisheries in India

Wildlife trade refers to the sale and exchange of animal and plant resources. The Wildlife Protection Act of 1972 is a package of legislation enacted in 1972 by Govt. of India. Wildlife (Protection) Act of 1972 accords protection to all forms of biodiversity, focused largely on terrestrial species. It was only in 2001 through a gazette notification that species such as marine sharks, rays and 15 kinds of molluscs were included after active campaigning by biodiversity experts.

The act consists of established schedules that lists protected plant and animal species. It has six schedules which gives varying degrees of protection. Schedule I and part II of Schedule II provide absolute protection-offences under these have prescribed the highest penalties. Species listed in Schedule IV are also protected, but the penalties are much lower. Enforcement authorities have the power to compound offences under this Schedule.

Quarantine

Quarantine means keeping the newly arrived animal or group of animals or plants in isolation for observations without any direct or indirect contact with other animals so as to prevent the spread of infectious pathogens and those animal are treated if necessary. Newly introduced animals could be carriers of diseases even if the animals appear to be in good health. The proper quarantine procedure protects the fish in the farm from being infected by newly introduced animals. Similarly, before shipment the fish intended for sale from the farm must be reasonably ensured that they do not carry any contagious disease. All live fish imports, including marines, are subject to quarantine and inspection before being released for sale. An ornamental fish which reaches to a hobbyist could be of either wild or farm rose. In the process, it passes through different categories of people like-collector, breeder, farmer, broker, exporter, wholesaler, retailer before it reaches to a hobbyist. As a fish moves from one source to another in the supply chain, it is either stocked in a new-holding facility for a varied period of time or also re-packed or both that result repeated and quick change of environment and fish get stressed.

Fig. 121: Harvesting of fish for sale

It is observed that in many instances fish may die because of stress only or it becomes susceptible to infestation by other pathogens initially and subsequently, mortality may occur if not controlled. It is estimated that about 50% of fish dies in Indian domestic trade during different stages of transit. Hence, there is need for acclimatization of fish as and when it is introduced to a new environment.

Fig. 122: Acclimatization of fish

Quarantine regulations

- Quarantine premises are privately owned but government-controlled.
- Consignments are only permitted from approved countries.
- A veterinary health certificate must accompany each consignment.
- All consignments must be inspected at the point of entry.
- All aquatic life other than approved fish is strictly prohibited.
- Fish are transferred by net from imported water into the fresh tank water.
- All imported water must be disinfected (chlorine or heat) before being disposed off into sewerage.

Quarantine Recommendations

Important suggestions to maintain quarantine protocol for freshwater ornamental fish should include the following:

1. It is always better to have a quarantine facility at hatchery/trade place.
2. Quarantine area should be isolated and separate from main facility.

3. Acclimatize the fish properly, which is the foremost requirement of quarantine.
4. Water quality parameters should not be only optimal but stable.
5. Quarantine tanks should have viewing facility that is adequate to observe the fish.
6. Each quarantine tank should have its own sets of equipment/filter system.
7. Standard prophylactic treatments should be carried out to reduce the stress and consequent incidence of diseases.
8. Bottom of the tanks should be bare, without sand, plants or anything.
9. Only authorized persons should be allowed to enter the quarantine area.

Quarantine period should be time-specific. A period of less duration is considered ineffective whereas a longer period is undesirable as well as uneconomical. The duration of quarantine may vary from species to species. Ideally, tropical fish should be quarantined at 22-25 °C and coldwater fishes at not less than 12-15 °C.

Fig. 123: Quarantine area

Table 11. Period of quarantine for different species

Gold fish	Gouramis & Cichlids	Other freshwater fin fishes	Fishes brought for breeding purpose
21 days	14 days	7 days	15-30 days (depending upon the species)

The application of various prophylactic treatments acts as preventive measures. Some of the commonly used prophylactic treatments for fish include dip bath or prolonged immersion in common salt, formalin, potassium permanganate, acriflavine and hydrogen peroxide.

Table 12. Commonly used chemicals for prophylactic treatment

Prophylactic solutions	Process
Common salt	Only un-iodised and preferably rock salt shall be used. Freshwater fishes entering quarantine should be given a saltwater dip (Sodium chloride crystals 5 g/l) if feasible, two more saltwater baths at 3-5 day intervals.
Formalin	Dissolve 1 ml formaldehyde in 10 litre of water and give an immersion treatment for about 1 hr. Formaldehyde could easily be obtained from a supplier of laboratory chemicals. Use of formaldehyde solution which appears milky should be avoided.
Potassium permanganate ($KMnO_4$)	Dissolve 4 gm $KMnO_4$ in1000 litre of water and give immersion treatment for 1-3 hrs. In case of prolonged treatment for 24 hrs. quantity of $KMnO_4$, is reduced to 2.5 g per 1000 litres.
Acriflavine	Dissolve 500 mg of acriflavine in 1 litre of water and keep it as a stock solution. Stock solution should be diluted and used as per requirement.

Biosecurity protocol

Biosecurity protocol is the practices and procedures used to prevent the introduction, emergence, spread, and persistence of infectious agents and diseases within and around fish production and holding facilities. It is also responsible in limiting conditions that can enhance disease susceptibility among the fish. It means biosecurity precautions are frame to exclude and contain fish pathogens. Biosecurity practices are applicable to all the levels of ornamental fish industry, starting from producers, wholesalers, retailered to the hobbyists. Proper use of biosecurity measures will help to prevent the introduction of infectious diseases in a fish culture facility, and will also help to minimize the risk of diseases being passed from producer to hobbyist in the chain.

The role of a veterinarian is becoming important to assist ornamental fish facilities in planning and implementation of biosecurity programs because now-a-days import-export regulations for ornamental fish are becoming increasingly stringent on global level.

Basically, biosecurity procedures are uniform across the industry, but the plan to implement it can be modified accordingly to meet the special needs of each business. The facility manager may modify and adjust the biosecurity measures for the basic tenets of good biosecurity practices as the scope, needs, and finances of the business are changing frequently.

Designing and implementing biosecurity practices can be simplified if some basic themes: pathogen exclusion, pathogen containment, and best health practices are considered. Poorly regulated international trade in ornamental fishes poses risks to both biodiversity and economic activity via invasive alien species and exotic pathogens. Good biosecurity includes obtaining healthy stocks and optimizing their health and immunity through good husbandry (animal management); preventing, reducing, or eliminating disease-causing organisms and their spread (pathogen management); and educating and managing staff and visitors on good biosecurity practices (people management). The importance of pathogen exclusion is to prevent the entrance of an infectious agent into a ornamental unit, thereby preventing infection and possible disease in a group of fish. To accomplish this, one must recognize and understand the various routes by which an infectious agent can enter a fish tank or pond. This helps to plan defensive measures that will check that entry.

Fish-associated entry

An obvious route of entry of pathogens into an ornamental fish unit is via the introduction of new fish. These fish may be asymptomatic carriers of a pathogen, or may have frank disease. It can be very difficult to determine whether the fish received are healthy, and rarely a manager can be totally confident that the fish that he has received are in fact healthy. To help minimize opportunities for diseased fish to enter a system, owners/managers must scrutinize potential suppliers before fish are purchased or shipped.

Water-associated entry

The presence and persistence of pathogens in water makes this medium a potential source of pathogen entry into a fish breeding and culture system. Water supply is a one of the major considerations while designing a biosecurity program based upon pathogen exclusion.

Food-associated entry

Fish food can not only serve as a source of pathogens, but poor, contaminated or spoiled diets can affect the fish and make them more susceptible to infection by pathogens. Therefore, it must be ensured that good quality commercial diets are being provided for the basic nutritional requirements of ornamental fish, and that can't accelerate the growth of infectious agents. Selection of traders to purchase fish food is also important and one should consider reputation and history of service when selecting food suppliers. The food should be carefully inspected to confirm that there is no spoilage. Live foods have higher chances to harbour pathogens. Therefore, it is advised to give more attention while testing the pathogens. Pretreatment or quarantine of the live food animals may be considered.

Person-associated entry

The people that enter a rearing system, whether staff or customers, should be considered in a biosecurity plan as they can be a source of pathogen introduction as well as pathogen persistence. Obviously, these people cannot be excluded from the facility, but the risks they pose can be managed.

Regulations impact on the ornamental fish and trade mainly related to the fish health, animal welfare, transport, phyto-sanitation of aquatic plants, and sustainable resource management. Fish health is one of the key regulations governing the trade through recently updated Aquaculture Health Directive of EU. This regulation is concerned to the animal health status and certification requirements for imports of tropical and cold-water fish for ornamental purpose. This EU directive implements to the OIE (Office International des Epizooties) guidelines and requires compliance by all EU members and their trading partners, including third countries. The directive is also linked to new transport regulations on the protection of live animals during transport and related operations. The imports of genetically modified organisms into the EU are also regulated by European Directives, governing the release and marketing of Genetically Modified Organisms (GMOs) in the European Union. Import of GMOs is currently banned in EU and Singapore but trading is allowed in countries such the USA and China. The green or eco-trend is also coming up in other leading importers of ornamental fish such as US.

Conservation and management strategy

Over 70 species of ornamental fishes have already been described from the Brahmaputra basin, of which at least 20 species can be reared in stagnant water without having much technical knowhow. Fish species belonging to the genera *Botia, Channa, Lepidocephalus, Somileptes* and *Trichogaster* are well known for their coloration. They are also available in low altitudinal streams. Certain other fishes like *Badis, Batasio, Glossogobius, Macrognathus, Mastacembelus, Olyra, Pseudoambassis, Rasbora* and *Tetraodon* can be regarded as potential ornamental fish. The species living in sluggish water and sandy bed habitat are mostly having silvery background with light coloured spots or stripes on lateral part of the body or orange tings on fins. The wetlands with weed infested muddy bottom are serving as refuge for multicolored species with varying body shapes. Habitat-wise distribution reveals that the hill stream fishes are streamlined, dorso-ventrally flattened, grayish or dark colored and most of them is bottom feeder, subsist mainly on benthos. One of the major problems with hill stream fish species is that they are very sensitive in terms of water quality and it is very difficult to rear them in captive condition. The maintenance cost to rear hill stream species is often very high and beyond the reach of ordinary people. Moreover, it has been found that a majority of the fish species, which were once common in the aquatic habitats of the northeast India, is highly depleted from the wild waters. Loss and degradation of habitat and unsustainable fishing as well as unauthorized collection are contributory factors for such depletion in their natural habitats. Recent survey of eight streams of Assam and Arunachal Pradesh indicated a drastic decline of the natural stock of the fish species belonging to genera like *Garra, Barilius, Batasio* and *Danio* that were fairly common in the upper Brahmaputra basin even about a decade ago. In addition to that, the current biodiversity status of potential ornamental fish species like *Channa barca* that was earlier enlisted as data deficient (DD) category should be reviewed. Similarly, emphasis need to be given to review the present status of species like *Botia dario, B. rostrata, Chaca chaca, Danio dangila, Rasbora rasbora, Badis badis, Erethistes pussilus, Gagatacenia*, etc., which were earlier included in the not evaluated (NE) category. Moreover, a few species can be presumed to be critically endangered from the wild waters of this region. Since tropical freshwater ornamental fish has a very high demand in the international market and considering the vast potentiality and commercial utilization of aesthetic fishes of the region and their effective conservation, there is a serious need for preparation of a database as well as re-evaluation of the status of aesthetic finfish of the region. The aforesaid problems associated with ornamental fish industry should be redressed at the earliest. For the purpose, a task force should be formed including all stakeholders associated with the industry. It should include experts, researchers, exporters, policy makers, small fish farmers, buyers as well as aquarists. Priority should be given to update the list of ornamental fish fauna of the region in the light of recent information and available data. Moreover, adoption of modern techniques for identification of fish species in order to get rid of taxonomic discrepancies is also suggested. Following strategies are to be adopted for the conservation of natural resources:

i) Restriction to net usage
ii) Conservation of breeding grounds
iii) Regulation of migratory fish may only secure the conservation
iv) Strengthening the indigenous fish population
v) Strict supervision and activation of cooperative society
vi) Proper management of data
vii) Awareness programme for local fishermen

The integration of all these requirements along with the community participation in resource conservation and sustainable use of the fish diversity is important.

Record keeping

Due to lack of information of what is happening at their farms, investors may take wrong decisions in absence of the records. The best sources of information needed to advise on the proper management of the ornamental fish farm are properly designed and maintained farm records. Record is the information that has been systematically and carefully collected and appropriately stored for intended use. In order to run any business successfully, it is important to collect information, document them properly and keep the records safely. For the purpose of keeping a track and decision making in the ornamental fish business, comprehensive and well-maintained records is a must. As in other enterprise, properly collected and kept records are also important in ornamental fish farming. The properly maintained records can be used to:

- Determine the profit of various techniques of production.
- Compare the efficiency of use of inputs, such as land, labour and capital, with that of alternative production activities.
- Help the investor in maintaining optimum efficiency of the farm's operations.
- Preservation of incidence as a memory of the enterprise for future reference.

  forecast the production quantity and value
- Plan the inputs required for a specific activity and estimate the fund requirement.
- Determine the financial health of the enterprise.

Farm record keeping methods

Farm record keeping methods range from the simple manual record (notebook & log book) to highly skilled digital computer accounting systems. The manual farm record book remains the old stand-by for farm record keeping due to its ease of use while the computer accounting systems vary in complexity and need for technical assistance. Many of the programmes available require some computer knowledge for fast and accurate management records. There are different types of farm record-keeping systems available in the market. For example, TALLY & QUICKEN computer software are widely used to keep financial records and Microsoft excel for input-output records. Review the different options before finalizing on the right system for your operation and decide for one that fits your specific farm operation. It is important to keep farm records as simple as possible but to record all necessary details in order that the performance of the farm operation can be fully evaluated.

Procedure to maintain records

- Keep the record maintenance as simple as possible. If the record-keeping system is unnecessarily complicated, there is more chance to make mistakes.
- To identify the facilities and fish stock (broodstock), it is essential to keep the identification records.
- Detail of all inputs and harvest from each unit should be recorded to estimate the profitability.
- To evaluate the breeding efficiency of farm broodstock detailed record of breeding indices (species wise & variety wise fertilization, hatching and survival rates) be recorded carefully.
- Record the details of feeds used in nursery pond – type and quantity, water quality in hatching and, nursery tanks/ponds.
- Stocking density in the culture system- species & variety wise, rearing duration & harvesting details with the value of harvested seed, etc. be recorded in a logbook for each and every cycle. Carry out a simple health inspection routine every day or weekly. If disease symptoms are detected, seek assistance from the expert and record all information in record logbook.

The records of the costs and income related to the ornamental fish business can be kept for cash analysis and enterprise appraisal. The most useful records are simple overview over the cash flow, that is, the total benefit in the farming.

Points to remember

- Water quality is always the central or contributing factor in disease outbreaks.
- Bacterial and parasitic diseases account for the majority of ornamental fish disease problems.
- Prevention is Always better than treatment.

Signs to be alert

- Excessive accumulation of debris, uneaten food, faces, and other muck in the aquarium or pond.
- Sudden changes in water quality parameters (ammonia, nitrite, pH, turbidity, alkalinity, dissolved oxygen, cloudy water, an unusual amount of foam in the system, etc.)
- A dead animal or an unusual mortality among animals that traditionally do very well in aquariums or pond environments.
- Distressed animals in the system. This is usually indicated by unusual activity, sluggishness, or unusual behaviour.

Common mistakes by an aquarist

- No knowledge about biology and husbandry requirements.
- Lack of information on quarantine new animals.
- Overcrowding the system. This will deteriorate the oxygen supply and the filtration systems, increase traumatic injury, territorialism, and cannibalism.
- Overfeeding.
- Not rinse adequately, rinse the recently cleaned and disinfected tanks, toys and equipment.
- Failure to quarantine newly arrived animals or to isolate those undergoing treatment.
- Initiating a disease treatment without proper diagnosis.
- Not to rinse the dust from activated carbon, dolomite or crushed shell before adding it to filters.
- Large aquariums not supported with well strong stands.
- Not understanding the importance of bypass and overflow pipes and screened drains.
- Use of copper, brass, or bronze valves and/or pipes. These can corrode, slough or leach toxic copper salts. Copper is especially toxic to invertebrates. Zinc is also quite toxic.
- Failure to provide proper substrates, shelter, or life support for animals.
- Failure to provide proper water flow and current for sessile invertebrates.
- Using plastics or sealers impregnated with insecticides or fungicides. Always read the label and when in doubt use food-grade containers.
- Using toxins or solvents in or around aquariums and ponds (insecticides, herbicides, floor strippers, cleaners, even smoke).
- Failure to keep certain species separate (predators with prey species, aggressive species with timid species, introducing parasites with host species).
- Failure to identify the individuals responsible for the care and maintenance of the animals and systems. Miscommunication can leave important husbandry tasks undone.
- Failure to check the water quality parameters regularly.
- Failure to observe and respond to declining water quality conditions.

- Inadequate nutrition due to underfeeding or an unbalanced or inappropriate diet. A varied diet is always best. Mono diets are never balanced.
- Failure to keep adequate husbandry records (water quality, feeding, mortalities, disease, and other significant events) and failure to review those records on a regular basis.
- Failure to recognize or anticipate the onset or duration of reproductive activity. Misinterpreting reproductive activity as abnormal behavior.
- Improperly installed/maintained electrical equipment and outlets not protected by ground fault interruption.
- Failure to check pipes, fittings and equipment on the suction side of pumps for air leaks. Air supersaturated water can kill animals quickly.
- Believing that antibiotics will solve all the problems.

13

Marketing and Trade

Ornamental fish culture has made giant strides in many countries in recent years. The industry is growing in leaps and bounds with involvement of even increasing number of aquarium hobbyists. A lucrative export market and high domestic demand have made ornamental fish industry a potential source for income generation. The range of species of ornamental fish available to the hobbyists is very large, with estimates of up to 1,500 species. As popularity of ornamental fish keeping has gained momentum, the need to transfer ornamental fishes from the resource abundant places to resource deficient places resulted in the marketing of ornamental fishes. In order to assess the marketing opportunities for ornamental fish, it can be classified as:

1. Local market
2. Inter state market
3. Export Market

Business Groups Associated with Aquarium Fish Trade

Breeders

Breeders are those, who rear or culture brood fishes of different species and produce fry. They sell fry of very small size to those who rear those fry and raise to larger sizes or to those who rear them on commission basis. Such people, sometimes, also rear those fry till they attain marketable size. Such people are very few in numbers.

Growers

Growers are those who buy small fry or procure on commission basis and rear them for 2 to 3 months till it becomes of considerably good size to fetch good market prices. Such people are more in numbers.

Carriers

They are the one, who buy fry or fishes from breeders and/or growers and sell them to growers or retailers in the markets.

Retail shops or Hobby centers

They maintain link among the above and some who buy fishes for hobby. They buy big fishes from the market or from the rearers and sell to hobbyist who possess aquaria. These people also sell aquaria and other related accessoires required for maintenance of aquaria. Such hobby centers or retail shops have important role in the development of such trade. They also render advice/ suggestions on different aspects in aquarium maintenance including water quality and feed management as well as control of diseases etc.

Wholesalers

They stock fishes of marketable sizes in large quantity and supply those to local buyers and also to other states. These people raise the fishes either themselves or procure through the growers on commission basis or buy from small parties (growers). They make maximum profit by putting least labour.

Exporters

A group of businessmen, who are involved in selling the selected or particular fish species or varieties of ornamental fishes to international markets, are termed as exporters.

Fig. 124: Weekly Domestic ornamental fish market of Kolkata

Ancillary/Supporting Unit

Apart from selling or trading of varied and beautiful fishes, many other related accessories required in maintenance of aquaria are also in the trade list. Those are aerators, air stones, filters, feeds (both live and dry), medicines, aquarium toys and aquatic plants etc. All this have pivotal role in aquaria or fish keeping hobby. Aquarium plants have good market demand. There are many aquarium plants those can be easily propagated and their culture can be undertaken on commercial scale. It fetches a very good price to an aquarist. Plants are the most beautiful and naturally available decorative materials. They provide beautiful background and natural environment to add aesthetic value in the aquarium.

Fig. 125: Marketing of fish and aquarium accessories

Internal Market in India

The turnover of domestic market is about Rs. 550 crore. This trade is practically restricted to limited merchants who, in fact, control and govern the trade. Kolkata ranks first in terms of overall turnover. The city and its suburbs have significant contributions in domestic market of aquarium fish trade as well as ornamental fish farming. Altogether there are 2200 heads in twin cities i.e. Howrah and Kolkata. Kolkata airport is the gateway for export of ornamental fish while on the other hand Howrah railway station is very important as the transit point for sending consignments of such pet fishes by trains to distant places like Delhi, Jaipur, Ahmedabad, Chennai, Mumbai and Bhopal. Such consignments of aquarium fishes, sometimes, are also airlifted to Bangalore, Chennai, Delhi, Dehra Dun and Ludhiana etc.

Economics of ornamental fish breeding and rearing units

A small-scale breeding/rearing unit earned an annual profit of Rs. 59,875 while medium scale breeding/rearing unit earned net profit of Rs. 1,66,550 per year for an initial investment of Rs 3,03000. In contrast, large farms with initial investment of Rs 8,00,000 earned very high net profit to the tune of Rs 4,12,000 per year due to large and modern infrastructural facilities. Rearing of exotic ornamental fishes fetches higher and steady returns than collection of fishes from wild due to their better quality and lower risk of mortality during transport. The small-scale retail outlets can earn an annual profit of Rs 16,400 for a total initial investment of Rs 24,000. In comparison to that, large-scale retail outlets are earning higher profits to the tune of Rs 36,200 per annum for an initial investment of Rs 92,000. Wholesalers are earning comparatively higher annual profit than the other stakeholders due to moderate initial investment and also due to the comparatively lower risk involved. In a case study, the small scale and large-scale wholesalers earned net profit to the tune of Rs. 1,91,100 and Rs 3,06,500 for initial investment of Rs 66,000 and Rs 1,70,000, respectively.

Table 13. Economic analysis of ornamental fish breeding and rearing

Particular	Small scale	Medium scale	Large scale
A. Initial Investment			
1. Land with compound wall	30,000 (46 sq. m.)	60,000 (92 sq. m.)	1,80,000 (139 sq. m.)
2. Office, storage, Electrical, Aeration filtration unit	8,000 (9 sq. m.)	16,000 (18 sq. m.)	40,000 (45 sq. m.)
3. Tanks (cement)	75,000 (13 sq. m.)	1,75,000 (32 sq. m.)	5,00,000 (46 sq. m.)
4. Small glass tanks	4,500 (6 sq. m.)	12,000 (18 sq. m.)	30,000 (45 sq. m.)
5. Other facilities	30,000	40,000	50,000
Total Investment	**1,47,500**	**3,03,000**	**8,00,000**
B. Operating cost (Electricity, water, fish and feed)	**50,000**	**70,000**	**1,00,000**
C. Depreciation @15%	22,125	45,450	1,20,000
D. Salary (calculated for one staff) @1500/ month	18,000	18,000	18,000
Total Fixed Cost	**40,125**	**63,450**	**1,38,000**
Total cost (B+C)	90,125	1,33,450	2,38,000
Annual sales	1,50,000 (1,00000 pieces)	3,00,000 (2,00000 pieces)	6,50,000 (5,00,000 pieces)
Net Profit	**59,875**	**1,66,550**	**4,12,000**

Table 14. Economics of ornamental fish retail outlets

Factors	Small scale	Large scale
A. Initial Investment		
1. Shops, storage, electrical, aeration filtration unit	8,000 (9 sq. m.)	40,000 (70 sq. m.)
2. Small glass tanks (23 sq.m.)	9,000 (40 sq. m.)	42,000
3. Packing facilities	7,000	10,000
Total Investment	**24,000**	**92,000**
B. Operating cost (Electricity, water, fish and feed)	**40,000**	**1,00,000**
C. Fixed cost Depreciation @15%	3,600	13,800
Total cost (B+C)	43,600	1,13,800
Annual sales	60,000 (15,000 pieces)	1,50,000 (40,000 pieces)
Net Profit	**16,400**	**36,200**

Table 15. Economics of ornamental fish wholesale farms in West Bengal

Factors	Small scale	Large scale
A. Initial investment		
1. Office, storage, electrical, aeration filtration unit	10,000 (18 sq. m.)	40,000 (46 sq. m.)
2. Small glass tanks (23 sq.m.)	6,000 (9 sq. m.)	30,000 (23 sq. m.)
3. Quarantine and Other facilities	50,000	1,00,000
Total Investment	**66,000**	**1,70,000**
B. Operating cost (Electricity, water, fish and feed)	50,000	1,00,000
C. Depreciation @15%	9,900	25,500
D. Salary (calculated forone staff) @1500/ month	18,000	18,000
Total operating Cost	**27,900**	**43,500**
E. Total cost (B+C)	**77,900**	**1,43,500**
F. Annual sales	97,000 (10,000 pieces)	4,50,000 (50,000 pieces)
G. Net Profit	**1,91,100**	**3,06,500**

Source: De and Ramachandran (2011)

Global market

Though the ornamental fish market's contribution to the world trade in terms of value is small, the sector plays a very important role in terms of poverty alleviation in developing countries. Coastal and riverine communities are able to utilize ornamental fish as a source of income. Global ornamental fish market crossed US $ 4.2 billion in 2017 and is projected to grow with a CAGR of more than 7.85%, in value terms, during 2019-2024 to reach over US $ 6.2 billion by 2024. Due to increasing prohibitions on keeping pets such as dogs and cats in high rise apartments, aquariums have become an important feature of the home décor, introduction and breeding of exotic species and growing interest in fish keeping has been driving the sale growth of this segment and expected to propel demand for ornamental fish over the next five years.

Initially, ornamental fish keeping was practiced in developed countries but it is also gaining attraction in developing countries as they have two third share of the total export value. The factor behind the expansion of ornamental fish market size is growing interest for aquarium fishes. This has expanded the trade activity of this particular market in more than 125 countries. With the growing popularity of household aquariums, the graph for global ornamental fish belonging to public aquaria sector is less than 1%. The import market is boosted in emerging developing countries of tropical and sub-tropical regions as most of the ornamental fish is sourced from there. The practice of international trade in ornamental fish consequently became the reason for employment opportunities to thousands of rural people in developing countries. The other contributing factors that propel the market growth includes advancements in transport, breeding, and keeping technology. However, supply issues can also be related to mortality caused by poor handling, destructive fishing methods, and quarantine procedures, and other man made factors like over-exploitation of a particular population of a species may determine the global ornamental fish market.

Export value for the Indian ornamental fish industry in 2016 was US$ 1.06 million and it contributed 0.3% of the total export. India ranks 31st in the world exporting countries list. Singapore, USA, Hong Kong, Malaysia and Japan were India's favourite top five market destinations.

Top exporting countries in the world

Singapore has been the ornamental fish capital of the world with an export value of US$ 42.97 million, contributing to 12.7% of the total exports. Till today, it remains the main trading hub in Asia, with more than 30% of the exported fish having been sourced from other countries. The second position is occupied by the Spain with exports of worth US$ 39.56 million, followed by Japan US$ 33.10 million, Myanmar US$ 32.05 million, Indonesia US$ 24.64 million, Czech Republic US$ 19.89 million, Malaysia US$ 14.09 million, Netherlands US$ 13.16 million, Sri Lanka US $12.61 million and Colombia US$ 10.68 million.

Top importing countries in the world

Among the top 10 importing countries USA was the single largest importer of ornamental fish with an import value of US$ 56.57 million, contributing to 19.7% of the total imports in 2016. The UK occupies the second position with imports worth US$ 23.02 million, followed by Germany at US$ 18.61 million, Japan US$ 15.98 million, Netherlands US$ 14.83 million, Singapore US$ 13.58 million, China US$ 12.62 million, France US$ 12.52 million, Hong Kong US$ 10.70 million, and Italy US$ 9.06 million. These 10 countries together shared over 83% of the global imports. Singapore, Germany (Frankfurt), Hong Kong, Malaysia and the Netherlands (Amsterdam) are the important trading hubs, re-exporting a major portion of their imports to these countries.

Important ornamental fish in global market

The global ornamental fish market has different varieties of fish; many of which are difficult to breed and expensive. Angel fish and gold fish are the most sought-after ornamental fish globally. The factors that have boosted ornamental fish market, especially of angel fish and gold fish, are their lovely appearances. They are native to Eastern Asia. The endemic ornamental species are dominantly of South American origin, but other regions neither realized its potential nor promoted it properly. Top brands are the gold fish, guppy, molly, platy, cichlid, tetra, catfish, gourami, loach, and the barbs. Despite being the country of 30 to 35 species of aquarist's favourite species, among the Asian origin, only *Pethia conchonius* and *Brachydanio rerio* are the most common fish, being exported. Hill stream fishes, belonging to the genera *Barilius, Homaloptera, Balitora, Garra, Nemacheilus* and *Lepidocephalus* are the species of coldwater but also found in warm waters. Their multiplication is easily achievable in stagnant water conditions of the aquaria. Thus, the market is prevalent.

World's most expensive tropical fish (Price in US)

1. Platinum Arowana - 400,000
2. Freshwater Polka Dot Stingray - 100,000
3. Peppermint Angelfish - 30,000
4. Bladefin Basslet - 10,000
5. Golden Basslet - 8,000
6. Neptune Grouper - 6,000
7. Australian Flathead Perch - 5,000
8. Wrought Iron Butterfly Fish - 2,700
9. Clarion Angelfish - 2,500
10. Candy Basslet - 1,000

World's most beautiful ornamental fish

1. Mandarin Dragonet
2. Juvenile Emperor Angel Fish
3. Lion Fish
4. Clown Trigger Fish
5. Nudi branch
6. Leaf Scorpion
7. Black Clownfish
8. Pink Spot Shrimp Goby
9. Blue Tang
10. *Acanthurus olivaceus*

World's most beautiful freshwater ornamental fish

1. Discus
2. Killifish
3. Male Betta
4. German Blue Ram
5. Endlers Livebearer
6. Boeseman's Rainbow fish
7. Gourami
8. Peacock Cichlid
9. Fantail Guppy
10. Flowerhorn

Areas of concerns

Problems with regard to supply, traceability and sustainable management along with the supply chain, disease, innovation as well as transportation practices, destructive fishing methods, and introduction of exotic species are some of the most important issues with regard to improving access to markets and enhancing value. With regard to supply, increasingly erratic climatic conditions and seasonal weather patterns, prevailing in Asia and South America, result in some species not being available during certain periods of the year. Supply problems are also related to over-exploitation of a species or a particular population of a species, destructive fishing methods, and mortality caused by poor handling and quarantine procedures, and other man made factors. This raises the issue of sustainability where there is a need for a healthy balance between the wishes of fish keepers, the economic interests of the business sector and the future of the species. The over-intensification of fish breeding as practiced in some countries of Asia has led to some serious problems such as susceptibility to disease, antibiotic resistance and poor brood stock quality affecting the trade.

Even though there is a good demand for Indian indigenous ornamental fish in the international markets, limited numbers are exported due to many reasons. The most important of these is the sustainability factor; secondly, there is not much interest in breeding indigenous fishes which are not popular in the domestic market. Although breeding techniques for selected indigenous ornamental fishes have been scientifically perfected in the country, their large-scale production is yet to begin. If the government institutions could set up large scale facilities and provide specialized training and assistance to the breeders, more indigenous ornamental fish can be produced for enhancing export from the country and earning foreign exchange.

Institutional Support

The economics and profitability of an ornamental fish-exporting unit works out to be highly lucrative, provided the activity is taken up on scientific lines with appropriate marketing strategies. The activity is possible not only on large scale but also on small scale. It provides good opportunity even to small entrepreneurs to enter into the field of ornamental fish trade. Institutional funding for R&D activities is, however, essential. Commercial banks can formulate schemes for extending financial assistance to prospective entrepreneurs for short-term training programmes on production of ornamental fishes. Two of the major areas which require immediate attention are (a) in-house breeding of selected species of ornamental fishes which are in great demand to release the pressure on wild capture and (b) scheme of educating/training a good number of fisher folk in the more skilled and specialized techniques of collecting, handling, storing and transport of ornamental fish which could revolutionize the fishery industry to some extent. It provides enhanced cash income and thus better living standards for those involved in the trade. The research institutions have been engaged in conducting educational courses as a part of their curricula to teach students as well as train farmers and entrepreneurs in aquarium fish culture, breeding of ornamental fish and their trade etc. The Department of Science & Technology (DST) has also come forward to support the NGOs to take up scientific breeding and culture of ornamental fishes in plains of North India and hilly areas of Uttrakhand and Sikkim. Several units, involving women with self-employment for income generation, are functional in the Himalayan hilly areas of Uttar Pradesh and Uttarakhand. It could be a formidable component for vocational training and income generating activities in rural sector as well.

For financial schemes of the agencies mentioned below, kindly refer details in the websites for schemes from time to time:

ICAR	www.icar.org.in
NFDB	www.nfdb.gov.in
MPEDA	www.mpeda.gov.in
NABARD	www.nabard.org

Developmental Schemes

Marine Products Export Development Agency. (MPEDA)

1. Scheme for providing financial assistance for establishment of Ornamental Fish Breeding Units: Salient features of the scheme are listed below:

- Generation of export-oriented employment in rural and urban households through mass production of ornamental fishes.
- 50% of the cost of eligible expenditure allowed, subject to a maximum of Rs. 75,000 for Grade I unit, Rs. 2 lakhs for Grade II unit and Rs 7.5 lakh for Grade III unit.

2. Scheme for providing financial assistance for establishment of Ornamental Fish Marketing Societies (OFMS)

To provide marketing infrastructure and to reduce intermediaries for ornamental fish breeders. The maximum amount of financial assistance eligible is Rupees Five lakh per unit.

3. Assistance for setting up of small-scale ornamental fish breeding units

To set up small scale ornamental fish breeding unit to enhance the domestic production. 50% of the capital cost for the infrastructure specified, subject to a maximum of Rs.75000/- per unit linked to export.

4. Developmental assistance for export of ornamental/aquarium fish

To give a thrust to the export of Ornamental fish from the country and to fetch better unit value realization by competing with our neighbouring countries. 10% of f.o.b. value of export subject to the ceiling of Rs.3 lakh per exporter per year.

National Agriculture Bank for and Rural Development (NABARD)

Scheme for Strengthening of Agricultural Marketing Infrastructure, Grading and Standardization

a) 50% of the subsidy amount will be released to NABARD by Department of Agriculture and Co-operattion in advance. Accordingly, NABARD releases subsidy to the participating banks in advance for keeping the same in a Subsidy Reserve Fund Account of the concerned borrowers, to be adjusted finally against loan amount of the bank on completion of the project.

b) The remaining 50% of the subsidy amount would be disbursed to the participating bank(s) by NABARD after a Joint Inspecting Committee comprising of officers from NABARD, participating bank and Directorate of Marketing &Inspection (DMI) in the concerned State, conducts an inspection.

National Fisheries Development Boand (NFDB)

Inland ornamental fishes and ornamental plants- breeding/rearing/ fabrication of aquariums.

Components of assistance

I. Setting up of Back-yard unit

II. Large scale unit

Eligibility criteria

1. Back-yard unit: Any house hold, SHG members who have sufficient back-yard area, and availability
2. Large scale unit: Entrepreneurs, Coop. Societies, Corporations, Federations and State Governments having at least 100 cents land with potential water source
3. Willing to take up the activity in accordance with the guidelines of NFDB
4. Prospective beneficiaries willing to undergo training

Unit cost (investment cost): The un it cost for establishment of back-yard hatchery is estimated at Rs.1,00,000/ and that of establishment of large scale unit in an area of 1000 sq. mts. is estimated at 8.00 lakhs. Farmers who wish to avail bank loan or who wish to invest on his own be provided with a subsidy not exceeding 20 per cent of the unit cost for establishment of large scale ornamental fishery unit while 50% grant would be extended to women/unemployed graduates for establishing backyard ornamental hatchery unit.

Cost of machinery for equipment for production of 5-9 t/h high quality floating fish feed: Rs. 464.20 lakhs

Construction of building, electrification, machinery erection, effluent system, etc.: Rs. 35.80 lakh ———————— Total: Rs. 500.00 lakh

Operation cost for first month (for production of 1200 tons/month and production cost @ Rs16/kg.): Rs. 19.00 lakh ———————— Total: Rs. 692.00 lakh

Table 16. Economics of Small Scale/Commercial Scale production of Ornamental fishes (Present prices)

S. No.	Item		Amount in Rs.
Capital cost			
1.	Low cost shed of 100 sft (Bamboo frame with net covering)	:	5,000
2.	Breeding tanks (6x3x2') 6 Nos	:	30,000
3.	Brood stock tanks. Circular (3* dia) 20 Nos	:	20,000
4.	Bore well (shallow) and pump & accessories	:	30,000
5.	Oxygen cylinder	:	5,000
	Total capital cost	**:**	**90,000**
Recurring costs (for 3 cycles)			
6.	Brood stock 100 pairs @ Rs.30/- pair (100 male and 100 female per cycle)	:	3,000
7.	Feed and medicines, per cycle	:	3,000
8.	Electricity charges, per cycle	:	1,000
9.	Breeding baskets & nets	:	1,000
10.	Miscellaneous	:	2,000
	Total recurring costs	**:**	**10,000**
Grand total		**:**	**1,00,000**
Gross Income per cycle of two months period			
No. of pairs		:	100
40% Response		:	40 pairs
Average .production per pair		:	100Nos
Total production		:	4,000Nos
80% survival		:	3,200nos
Nursery gate price @Rs.4/- piece		:	1,28,000/-
Income on 3 cycles		:	38,400/-
Total income per year (6 cycles) 12200x6		**:**	**76,800/- or 77,000/-**
Total cost			
10% depreciation on capital cost (Rs.90,000)		:	9,000
Interest on the loan (Rs.50,000)@12% p.a.		:	5,000
Recurring cost for 6 cycles		:	20,000
Miscellaneous		:	35,000
Net income per year 77,000 - 35,000		**:**	**42,000**

NFDB Schemes for funding ornamental fisheries

a) Backyard Hatchery of estimated cost Rs. 1.50 lakh a subsidy of 40% of unit cost to members of Women SHGs/ Fisherwomen Cooperative Societies/ Entrepreneurs.

b) Medium Scale Unit costing Rs 4.00 lakh, 40% of unit cost as subsidy to all categories of beneficiaries.

c) Integrated Ornamental Fishery Units, costing Rs 15.00 lakh, a subsidy of 40% to the Government Agencies/ Government Institutions/ Entrepreneur

d) For Setting up of Aquarium Fabrication Units of cost Rs. 1.00 lakh a subsidy of 40% per unit cost to members of Women SHGs/ Fisherwomen Cooperative Societies while 25% of per unit cost as subsidy to Entrepreneurs/individual persons.

Table 17. Pattern of assistance (NFDB, Hyderabad)

S.No.	Name of the Activity/Scheme	Unit Cost	Pattern of Assistance
a)	Backyard Hatchery	Rs. 1.50 lakh	40% unit cost as subsidy to members of Women SHGs/ Fisherwomen Cooperative Societies/ Entrepreneurs
b)	Medium Scale Unit	Rs 4.00 lakh	40% unit cost as subsidy to all categoriesof beneficiaries
c)	Integrated Ornamental Fishery Units	Rs 15.00 lakh	40% Subsidy to the Government Agencies/ Government Institutions/ Entrepreneur
d)	Setting up of Aquarium Fabrication Units	Rs. 1.00 lakh	40% unit cost as subsidy to members of Women SHGs/ Fisherwomen Cooperative Societies.25% unit cost as subsidy to Entrepreneurs/individual persons

Constraints

1. Conservation measures for protection of natural resources is lacking.
2. R & D (Scientific farming) support is not satisfactory and sufficient.
3. Pollution due to discharge of agricultural and domestic wastes in the rivers and canals is a serious threat to this industry.
4. Financial support is insufficient or lacking.
5. No co-ordination exists among traders, businessmen, farmers, scientists, state officials and financiers etc.
6. Proper and timely air flights connecting link between many places are lacking or insufficient.
7. Non-organization of awareness programmes about the merits and benefits of culture and breeding, trading and conservation etc. of ornamental fishes is a serious concern.
8. Strict vigilance or certification for trading of quality seed for export and import is lacking.

9. Road linkage with interior places is lacking.
10. Encouragement to bring the youth especially unemployed, educated is lacking to make them involved in such industry.
11. Publication on aquarium fishes as firsthand knowledge and information and about their benefitsetc. in zonal / regional languages is not available.
12. Formation of co-operative society to make the farmers involved in such activities is also at bay.
13. No strict quarantine and disease control measures are adopted in many cases.

The overall development, welfare and improvement etc. of such industry could be possible only after considering proper re-dressal of the above mentioned constraints.

Pradhan Mantri Matsya Sampada Yojana (PMMSY)

In the Year 2020 a scheme to build about Blue Revolution through sustainable and responsible developmental of Fisheries sector in India is initiated. For implementation of the PMMSY over the periods of Five year from 2020-21 to 2024-25 a total investment of Rs. 20,050 crores comprising (a) Central share of Rs. 9,407 crore (b) State share of Rs. 4,880 crore and (c) beneficiaries share of of Rs. 5,763 crore has been estimated. Out of total estimate of Rs. 20,050 crore the Ornamental fish trade is allocated a share of Rs. 576 crore.

Name of activities	Unit Cost (Rs. In lakh)	Approximate Physical Quantities	Total Cost (Rs. in Crore)
Development of Ornamental and Recreational Fisheries			
Backyard Ornamental fish Rearing unit (both Marine and Fresh water)	3	1010	30.30
Medium Scale Ornamental fish Rearing unit (Marine and Freshwater Fish)	8	707	56.56
Integrated Ornamental fish unit (breeding & rearing for freshwater fish)	25	404	101.00
Integrated Ornamental fish unit (breeding and rearing for marine fish)	30	303	90.90
Establishment of Freshwater Ornamental Fish Brood Bank	100	10	10.00
Promotion of Recreational Fisheries	DPR	DPR	25.00
Subtotal (A)			**313.76**

Technology infusion and adaptation			
Establishment of large RAS tank capacity (with 8 tanks of minimum 90 m3/	50	50	25
Establ ishment of Medium RAS (with 6 tank)	25	100	25
Establishment of small RAS	7.5	200	15
Establishment of Backyard mini RAS units	0.5	200	1
Live fish vending Centres	20	110	22
Fish Feed Miils(mini)	15	50	3
Markets and marketing infrastructure			
Construction of fish retail markets including ornamental fish/aquarium markets.	100	20	20
Construction offish kiosks including kiosks of aquarium/ornamenta! fish	10	200	20
E-platform for e-trading and e-marketing of ornamental fish	Proposal/DPR based	5	5
Innovative activities, Start-ups etc (10 lakh Gold fish for girl child)	DPR		20
Genetic improvement	DPR		10
Aqua park + Aquarium			100
Subtotal B			**261.00**
		Total (A+B)	**576.00**

Activities can be grouped as:

- Group A: Activities related to production of ornamental fish (Setting up of production units including renovation)
- Group B: Activities related to Aquarium Fabrication, trade and marketing
- Group C: Activities for promotion of ornamental fisheries sector through demonstration, establishment of public aquariums and organizing aquaria shows
- Group D: Skill development and capacity building Programmes. Potential for establishing aquaria in educational institutions

14

Frequently Asked Questions

1. Choose the most appropriate option with a tick mark

i) Floating plants are required for breeding

a) Guppies
b) Angel fish
c) Swordtail
d) Gourami

ii) Breeding traps are generally used for

a) Egg-scatters
b) Egg-depositors
c) Livebearers
d) Nest-builders

iii) Gonopodium is present in

a) Female live-bearer
b) Female egg-layer
c) Male live-bearer
d) Male egg-layer

iv) Scientific name of Dwarf Gouramy

a) *Trichogasterfasciata*
b) *Trichogasterchuna*
c) *Trichogasterlalius*
d) *Trichogasterlabiosa*

v) Example of an indigenous ornamental fish of Indian region

a) *Colisalalia*
b) *Carassius auratus*
c) *Trichogasterleeri*
d) *Pterophyllumscalare*

vi) Aquarium fish may excrete ammonia in the range of

a) 0.3 – 4.0 g/kg fish / day
b) 0.2-4.0g/kg fish /day
c) 1.0-5.0g/kg fish/day
d) None

vii) Protein Skimmer is an

a) Biological filtration method
b) Chemical filtration method
c) Mechanical filtration method
d) All of above

viii) The largest wholesale market of ornamental fish in the Eastern and North Eastern Zone of India

a) Kolkata b) Chennai

c) Kochi d) Visakhapatnam

ix) Thickness of glass for aquarium fabrication depends on

a) No. of fishes b) Type of fishes

c) Water volume d) Water Temperature

x) *Hyphessobryconpulchripinis* is

a) Gold Spotted Tetra b) Black Neon Tetra

c) Bleeding Heart Tetra d) Lemon Tetra

xi) Guidelines for the import of ornamental fishes into India is formulated by

a) MPEDA b) NFDB

c) DAHDF d) EIA

xii) Male possess sword like caudal fin in

a) *Xiphophorus variatus* b) *Xiphophorus hellerii*

c) *Xiphophorus maculatus* d) *Xiphophorus nezahualcoyotl*

xiii) Dropsy is caused by

a) *Aeromonas hydrophilla* b) *Pseudomonas punctate*

c) *Saprolegnia* d) a & b

xiv) Which of the following breed of goldfish has nacreous scales with mixture of different colors?

a) Comet b) Ryukin

c) Shubunkin d) Demekin

xv) Activated carbon removes the waste materials from aquariums based on the principles of

a) Absorption b) Adsorption

c) Diffusion d) Dilution

xvi) Quinaldine can be used during the transportation of ornamental fish as

a) Sedative drug b) Disinfectant

c) Antibiotic d) Probiotic

xvii) Flower horn is a

- a) Hybrid of cichlids
- b) Transgenic fish
- c) Breed developed through chromosome manipulation
- d) None of the above

xviii) Aquatic plants like hydrilla are placed in breeding tank of Goldfish

- a) To reduce the temperature in the breeding tank
- b) To collect the fertilized eggs of goldfish
- c) To provide oxygen to the eggs of goldfish
- d) None of these

xix) The newly born hatchlings of angelfish are attached to leaves by using the sticky thread attached to

- a) Head
- b) Vent
- c) Anal fin
- d) Pelvic fin

xx) The newly born hatchlings of angelfish are attached to leaves by using the sticky thread attached to

- a) Head
- b) Vent
- c) Anal fin
- d) Pelvic fin

xxi) Which of the following genetic improvement programme may require long time to produce quality strains of ornamental fish

- a) Selective breeding
- b) Transgenesis
- c) Androgenesis
- d) Gynogenesis

xxii) Which of the following is a motor driven mechanical filter?

- a) Under gravel filter
- b) Sponge Filter
- c) Box Filter
- d) Canister filter

xxiii) Select the right answer

1. Origin of ornamental fish culture
   - a) China
   - b) Japan
   - c) England
   - d) UAE
2. Modern aquarium fish keeping began in the year
   - a) 1990
   - b) 1760
   - c) 1805
   - d) 1899

3. First public display aquarium opened at Regent's Park in

a) 1993 b) 1960
c) 1853 d) 1800

4. One of the best ornamental fish keeping facilities in the world exists at Singapore, which is called

a) Aquarium b) Gallery
c) Pendalium d) Oceanarium

5. In India, the first public Aquaria in India was established in middle of the 20th century at

a) Vishakhapatnam b) Kolkata
c) Cochin d) Mumbai

6. Small fishes like *Botiadario, Daniodangila, Puntiusshalynius* and *Schistureticulofasciatus* which can be reared in aquarium throughout their life span are called

a) Young ornamentals b) Classified ornamental fishes
c) Coloured ornamentals d) Split ornamentals

7. Larger food fishes like *Neolissocheilushexagonolepis, Labeogonius, Channamarulius* and *Rita rita* which are treated as ornamental fish in their juvenile stages are called

a) Young ornamentals b) Non-Classified ornamental fishes
c) Coloured ornamentals d) Split ornamentals

8. Keeled abdomen is an ornamental characteristic that exists in the fish

a) Notopteruschitala b) Badisbadis
c) Chela laubuca d) Labeorohita

9. In India, this region is a biodiversity hotspot

a) Western Ghats b) Eastern Ghats
c) North India d) Karnataka

10. High priced fresh water native ornamental fish in India

a) Arowana b) Goldfish
c) Barca snakehead d) Angel fish

xxiv) Select right option

1. Example for dug-out ponds

a) Earthen ponds b) Glass tanks
c) Ferro cement tank d) All of the above

2. Scientific name of fighter fish

   a) *Xiphophorushelleri* b) *Pterophyllumscalare*

   c) *Betta spleridens* d) *Hyphessobrycon serape*

3. General size of aquarium

   a) 10" x 10" x 10" b) 60" x 18" x 18

   c) 18" x 12" x 12" d) All of the above

4. World's largest producer and importer of ornamental fishes

   a) USA b) Singapore

   c) India d) Japan

5. Capacity of the fibre tanks

   a) 200 L b) 100 L

   c) 50 L d) All the above

6. Contribution of freshwater ornamental fishes to Indian ornamental fish trade industry

   a) 30% b) 50%

   c) 65% d) 85%

7. As per FAO's report India is ranked .... in export value

   a) 20th b) 24th

   c) 30th d) 15th

8. Mazla is type of

   a) Earthern container b) Glass container

   c) Fiber container d) Cement container

9. Number of fishes valued by ornamental fishery trade of India from freshwaters

   a) 250 b) 100

   c) 500 d) 50

10. Common name of *Puntius tetrazona*

   a) Molly b) Guppy

   c) Barbs d) Gold fish

xxv) Which one is the right choice?

1. Ultra-plankton is referred to the planktons measuring

a) 0.5-10 ìm b) 10-50 ìm
c) 50-500 ìm d) >500 ìm

2. Euglenophyceae for fishes, is an important group of plankters as its source of

a) guanine b) food
c) Biodiesel d) Medicine

3. Spirogyra is an example of a green algae

a) Free floating
b) Multi cellular non filamentous
c) Multi cellular unbranched filamentous
d) Colony forming

4. Blue green algae belong to

a) Xanthophyceae b) Euglenophyceae
c) Chlorophyceae d) Cyanophyceae

5. Name "living capsules of nutrition"

a) Virus b) Bacteria
c) Blue green algae d) Zooplankton

6. Most primitive organism in animal kingdom is

a) Bacteria b) Infusoria
c) Virus d) Green Algae

7. Most popular rotifer used as fish feed is

a) *FIlinia* b) *Brachionus*
c) *Asplanchna* d) *Keratella*

8. Water fleas are common name for

a) Copepods b) Artemia
c) Cladocerans d) Infusoria

9. Ostracods are

a) Bivalve crustaceans b) Primitive algae
c) Toxic phytoplankton d) Important for biodiesel production

10. Brine shrimp are commonly known as
    a) Sea Lion b) Sea lettuce
    c) Sea Monkey d) Sea cowries
11. Institutes provide financial assistance for ornamental fish culture are
    a) NFDB b) MPEDA
    c) Both d) None
12. Quarantine period for gold fish is
    a) 100 days b) 50 days
    c) 20 days d) Not required
13. Bio security is essential for fish
    a) To avoid infections b) To enhance the color
    c) To grow the size d) To remove dead fish
14. To introduce the new variety of fish in the culture system
    a) Light must be on b) Light must be off
    c) Light does not matter d) It must be raining
15. License to export ornamental fish can be issued by
    a) Banks b) MPEDA
    c) University d) Police

2. State 'True' or 'False'

i) Platy is a live bearer
ii) *Osphronemus gourami* is an indigenous ornamental fish.
iii) Inter-specific hybridization is a method for mono-sex male production of ornamental fish.
iv) The Wildlife Protection Act of India consists of 8 established schedules which give varying degrees of protection.
v) Beta-carotene gives yellow coloration in fish.
vi) *Dactylogyrus* and *Gyrodactylus* are endoparasite.
vii) Activated carbon is most common type of biological filtration media.
viii) Gold fishis considered as egg scatterer and laying non-adhesive eggs.
ix) Commonly painted Indian glass fish is*Parambassislala*.
x) The Japanese started the selective breeding of Goldfish from the year 1000 AD.

xi) Fin and tail rot produced by Fungi.

xii) White spot or Ich disease in ornamental fish is caused due to virus

xiii) Artificial illumination by tungsten bulbs is better than fluorescent lamp.

xiv) Rasbora requires slightly acidic water for breeding.

xv) Gourami is a bubble nest builder for laying eggs.

xvi) Granular feeds have smaller surface area compared to powdered feeds.

xvii) Water quality and balanced diet are very important for aquarium fish keeping; however, the former is more critical on long-term basis and the latter on short-term basis.

xviii) Tropical warm-water fishes including tetras, zebras, cichlids and gouramis require feeds at 1-2% of their body weights per day.

xix) Revenue earned from sale of aquarium fish far exceeds revenue earned from aquarium and its accessories in domestic market.

xx) An aquarium tank should consist of a balanced system of animals, plants and bacteria for maintenance of water quality.

xxi) Leaching of nutrients from fish feeds mainly depends on the surface area of the feeds.

xxii) Gonopodium in live bearers is a modification of pelvic fin.

xxiii) Breeding tanks prepared for tiger barb need no planting.

xxiv) Breeding traps are associated with breeding of goldfish.

xxv) Ceratophyllum is a good example of an exotic ornamental plant.

3. Fill in the blanks with most appropriate word

i) ____________________ Ornamental fish family exhibit parental care.

ii) Two common non-steroidal aromatase inhibitors are ____________ and ____________________.

iii) ____________ and ____________ are an important cladoceransgroup of live food used in aquarium.

iv) Ammonia is converted by ____________________ bacteria to nitrite and further nitrites converted by ____________________ to nitrate.

v) ____________________ and ____________________ are causative agent of Cotton wool disease.

vi) Gonopodium is modification of ____________________.

vii) For providing oxygen to the developing eggs the angelfish uses __________ fin.

viii) The trade name of genetically modified to fluoresce in bright colors under white or ultraviolet light commercially available in the United States are known as ______________.

ix) Freshwater blood parrot cichlid fish is a hybrid species of __________ and ________________.

x) The Wildlife Protection Act of ________________________ refers to a package of legislation enacted by the Govt. of India applicable to entire India except ______________________.

xi) ______________________ containing food enhance the coloration in ornamental fish.

xii) The scientific name of Red line torpedo fish, an important indigenous fish of Kerala is ____________________

xiii) Water discoloration due to woods occur in aquarium due to higher contents of ____________________

xiv) ____________ is used as adhesive while fabricating whole glass aquarium.

xv) Dropsy in the ornamental fish is caused due to ______________________

xvi) *Artemia salina* commonly known as _______________

xvii) ___________ are useful for production of small air bubbles and for efficient oxygenation.

xviii) Live foods like tubifex and chironomid larvae can be preserved in the form of ___________________ for future use.

xix) Protein requirement of the young stages of ornamental fish is _________________

xx) ____________ is known as 'living nutritious capsule'

4. Short Answer Type

i) Construction and fabrication of glass aquarium

ii) Goal of selective breeding in ornamental fishes.

iii) Different types of aquarium filters used in aquarium

iv) What is ornamental fish?Criteria for being an ornamental fish with example.

v) Mouth-brooders

vi) Write short notes on

a) Community aquarium.

b) Breeding traps.

c) What should an aquarist do with his/her aquarium while going on a vacation?

d) Nest builders.

e) Types of aquatic plants used in aquarium.

f) Gonopodium.

vii) List the categories of air pumps used in aquaria.

viii) Explain methods for dealing with the problems of algal growth in an aquarium.

ix) How will you transport 500 nos. of 6 inches gold fish to a distance of 1000 kms?

x) Why do you think that quarantine of fish is required before bringing to the farm?

xi) Name five indigenous ornamental fish.

xii) Name five marine ornamental fish of India.

5. Long Answer Type

i) Write a brief note on guidelines of MPEDA for ornamental fish export.

ii) Explain breeding biology and seed production technique of Angel fish.

iii) Illustrate processes of production of monosex guppy by Letrozole.

iv) Describe different filtration methods in aquarium management.

v) Why do Water quality management in aquarium is important?

vi) Illustrate sexual dimorphism and breeding techniques of Tiger barb, *Puntius tetrazona*

vii) What is the role of Carotenoids in color enhancement in ornamental fish?

viii) Explain the packaging methods used in trade of ornamental fish

ix) Explain different designs and materials used for construction of aquarium. Describe the steps to be followed for construction and fabrication of a glass aquarium.

x) Write down the detail procedures for setting of a freshwater aquarium for domestic and breeding purpose.

xi) Define filtration. Explain the three filtration systems such as Mechanical, Chemical and biological with examples.

xii) Write down the principles of a balanced aquarium. Explain the role and mechanisms of biofiltration in ornamental fish farming system.

xiii) Explain the benefits of using aquatic plants for maintaining and farming of ornamental fish. Write down at least ten commonly used plants for freshwater tropical aquaria.

xiv) Describe the resources and trade potentiality of indigenous ornamental fish of India. Write down at least ten different indigenous ornamental fish available in different parts of India.

xv) Explain the broodstock management, breeding and rearing practices of Goldfish.

xvi) Explain the broodstock management, breeding and rearing practices of Livebearers.

xvii) Write down the different diseases commonly seen in ornamental fishes and their control measures.

xviii) Describe different genetic improvement programmes with examples for producing quality strains of ornamental fish.

xix) Describe the methods for controlling growth of algae in aquarium tanks.

xx) Draw a schematic diagram of an under-gravel filter in an aquarium. Mention the advantages and disadvantages of this filtration system.

xxi) Write about breeding biology and seed production technique of Gold fish.

xxii) Describe different filtration methods in aquarium management.

xxiii) Write a short note on the behavior and breeding techniques of fighter fish.

xxiv) Write about the ornamental fish trade in the world. Give an account of export trade of ornamental fishes from India.

xxv) Describe packaging methods used in trade of ornamental fishes.

xxvi) Color maintenance and color enhancement in ornamental fish.

xxvii) Define fish acclimatization and explain the process.

xxviii) What are the constraints in breeding the indigenous ornamental fish?

xxix) Highlight the potential and prospect of ornamental fish trade in India.

xxx) Explain the feed and feed management for brood stock rearing in an ornamental fish farm.

Suggested Readings

Abraham, T.J., Sasmal, D. and Banerjee, T., 2004. Bacterial flora associated with diseased fish and their antibiogram. *Journal of Indian Fisheries Association***31**: 177-180.

Ahilan, B. and Jeyaseelan, P., 2001. Effects of different pigment sources on colour changes and growth of juvenile *Carassiusaurtatus. J.Aqua. Trop.***16(1)**: 29-36.

Alagappan, M., Vijula, K. and Sinha, A., 2004. Utilization of spirulina algae as a source of carotenoid pigment for blue gouramis (*Trichogastertrichpterus* Pallas). *Journal of Aquariculture and Aquatic Sciences*, **X(1)**: 1-11.

Ali, M., Raizada, S., Maheshwari, U.K., Chadha, N.K., Javed, H. and Kumar, S., 2004. Effect of salinity on broodstock development and the spawning of blue gourami (*Trichogastertrichopterus*). *In:* National Seminar on Prospects of Ornamental Fish Breeding and Culture in Eastern and North-Eastern India, 16-17 February, 2004. pp. CA-42.

Bijukumar, A., 2000. Exotic fishes and freshwater fish diversity. Zoos' Print J., **15(11)**, 363–367.

Dafni A, Lev E, Beckmann S, Eichberger C. 2006. Ritual plants of Muslim graveyards in northern Israel. J Ethnobiol Ethnomed **2:38**. doi:10.1186/1746-4269-2-38

Das R.C. and Sinha A., 2003. Ornamental fish trade in India. *Fishing Chimes*, **23(2)** 16-18.

Dey, S., Ramanujam, S. N. and Mahapatra, B. K. 2014. Breeding and development of ornamental hill stream fish *Devarioaequipinnatus* (McClelland) in captivity. International Journal of Fisheries and Aquatic Studies. **1(4):** 01-07.

Dhert P. and Sorgeloos P. 1995. Live feeds in aquaculture. *In* "Aquaculture towards the 21 st century': Proceedings of the INFOFISH – AQUATECH 94 conference, Colombo, Sri Lanka, 29-31 August, 1994, 209-219.

Dudgeon D., 2003. The contribution of scientific information to the conservation and management of freshwater biodiversity in tropical Asia. *Hydrobiologia* **500:** 295–314.

Estrada-Castillón E, Garza-López M, Villarreal-Quintanilla JA, Salinas-Rodriguez MM, SotoMata BE, González-Rodriguez H, González-Uribe DU, Cantú-Silva I, Carrillo-Parra A, CantúAyala C (2014) Ethnobotany in Rayones, Nuevo León, México. J Ethnobiol Ethnomed **10**:62.

Froese R. and D. Pauly (eds.), 2016. FishBase. www.fishbase.org. Electronic version accessed 27/04/2020.

Ghosh S., Sinha, A., Yhome, V. and Sahu, C., 2005. Bacterial load associated with culture of ornamental fishes in glass aquarium. *Environment & Ecology*, **23(1):** 118-123.

INFISH, 2015. National Fisheries Development Board Newsletter, Volume 7, Issue 1 April - June: 3-5.

James, R. and Sampath, K., 2003. Effects of meal frequency on growth and reproduction in the ornamental red swordtail, *Xiphophorus helleri. The Israeli Journal of Aquaculture-Bamidgeh*, **55(3):** 197-207.

James, R. and Sampath, K., 2004. Effect of feeding frequency on growth and fecundity in an ornamental fish, Betta splendens (Regan). *The Israeli Journal of Aquaculture-Bamidgeh*, **56(2):** 138-147.

Jameson, J.D. and R. Santhanam, 1996. Manual of Sindermann and Lightner [20, 21]. Trout and salmon are ornamental fishes and farming technologies. attacked by Saprolegnia and Achlya spp. and catfish by Tuticorin, India: Fisheries College and Research Saprolegnia spp. Shrimp are infected by Lagenidium and Institute, pp: 1-4.

Jha, P., Sarkar, K. and Barat, S., 2004. Effect of different application rates of cow dung and poultry excreta on water quality and growth of ornamental carp, *Cyprinus carpio* var. koi, in concrete tanks. Contributed abstract, *In:* National Seminar on Prospects of Ornamental Fish Breeding and Culture in Eastern and North-Eastern India, 16-17 February, 2004. pp. CA-8.

Kirankumar, S. and Pandian, T. J., 2004. Production and progeny testing of androgenetic rosy barb *Puntius conchonius J. Exp. Zool.* **301A**:938-951. © 2004 Wiley-Liss, Inc.

Kirankumar, S. and Pandian, T.J., 2002.Effect on growth and reproduction of hormone immersed and masculinized fighting fish *Betta splendens J. Exp. Zool.* **293**:606-616 . © 2002 Wiley-Liss, Inc.

Kumbhar BA, Dabgar PK, 2014. To study of aesthetic values of some traditional worshiping plants of Dang District. Int J Sci Res **3(4)**:46–47

Li XX, Zhou ZK. 2005. Endemic wild ornamental plants from Northwestern Yunnan, China. Hortscience **40(6)**:1612–1619

Mahapatra BK and Gopal Krishna, 2016. Embryonic and larval development of *Rasbora daniconius* (Hamilton): A potential indigenous ornamental fish of north-east India, *International Journal of Fisheries and Aquatic Studies*; **4(6)**: 187-190.

Mahapatra, B. K. and Lakra, W. S., 2014. Biology of *Brachydanio rerio* (Hamilton, 1822) from NEH region, India. *J. Inland Fish. Soc.* India, **46(2)**:64-70, 2014.

Mahapatra, B.K.,Vinod, K. and Mandal, B.K., 2003. Prospects of *Puntius* (Barbs) in North Eastern Hill Region with reference to export potentiality in ornamental fish trade. Proc. Participatory Approach for Fish Biodiversity Conservation in North East India (Eds. P.C.Mahanta and L.K. Tyagi), National Bureau of Fish Genetic Resources, Lucknow, pp. 214-224.

Mercy T.V.A., Malika V. and Sajan S., 2013. A reproductive biology of *Puntius denisonii* (Day 1865) – an endemic ornamental cyprinid of the Western Ghats of India. Indian J. Fish., **60(2)**: 73-78.

Monica F. Solberg, Grethe Robertsen, Line E. Sundt-Hansen, Kjetil Hindar and Kevin A. Glover, 2020. Domestication leads to increased predation susceptibility, *Scientific Reports*, 10.1038/s41598-020-58661-9, **10**, (1):

Nirmal Kumar JI, Soni H, Kumar RN (2005) Aesthetic values of selected floral elements of Khatana and Waghai forests of Dangs, Western Ghats. Indian J Tradit Knowl **4(3)**:275–286

Oloyede FA., 2012. Survey of ornamental ferns, their morphology and uses for environmental protection, improvement and management. Ife J Sci **14(2)**:245–252

Pailan G.H., Sinha A. and Kumar M., 2012. Rose petal meal as a natural carotenoid source for pigmentation and growth of rosy barb (*Puntius conchonius*). *I.J. Anim. Nutr.* **29(3)**: 291-296

Pailan G.H., Sinha A. and Borkar A.,2012. Rose petal meal as a natural carotenoid source for pigmentation and growth of dwarf gourami *Colisa lalia*. *Ani. Nutr. Feed Tech.,* **12(2)**: 199-207

Pal, M. and Mahapatra, B.K. 2017. Early Life History of Indian Ornamental Barb, *Puntius sophore* (Hamilton, 1822). *J. Inland Fish. Soc. India*, **49(2)**:10-21.

Prem Kumar K.P. and N.K. Balasubramanian, 1984. Breeding biology of the scarlet banded barb, *Puntius amphibius*(Val.) from Chackai Canal. Zoologis cherAnzeiger Leipzig 213 **(3-4)**: 291-302.

Reddy A.K., 1997. Culture of live food organisms for ornamental fishes. *In*: "Advances in keeping and breeding ornamental fishes", Training Manual, CIFE, Mumbai, 34-41.

Reddy A.K. and Thakur N.K. 1997. Artemia - its use as live food in aqua hatcheries. Practical Manual, Aquaculure Division, CIFE, Mumbai, 26 pp.

Saini V.P., 2016. Record keeping at ornamental fish production and marketing establishments. In Jain A.K., Saini V.P. and Kaur V. I. Best management practices for freshwater ornamental fish production, NFDB, Hyderabad, India.153-164.

Sales J. and Janssens P.J., 2003.Nutrient requirements of ornamental fish-Review. *Aquatic Living Resources* **16**: 533–540.

Silas, E.G., Gopalakrishnan, A., Ramachandran, A., Anna Mercy, T.V., Kripan Sarkar, Pushpangadan, K.R., Anil Kumar, P., Ram Mohan, M.K. &Anikuttan, K.K., 2011. Guidelines for Green Certification of Freshwater Ornamental Fish. The Marine Products Export Development Authority, Kochi, India. xii + 106 p.

Sane S.R., 1982. Present status of the trade and future export prospects, In: Proceedings of the seminar on Prospects of aquarium fish export from India (10 February 1982, Bombay), MPEDA, Cochin: 25-36.

Sinha, A., Ghosh, S. and Singh, D., 2004. Probiotics as nutrient supplement in artificial feed of gold fish (*Carassius auratus*). *J. Indian Fish. Assoc*., **31**: 139-144.

Sinha A., 2004. Ornamental fishes. Published by Director, CIFE, Mumbai, 54p.

Ghosh S., Sinha A.,Yhome V. and Sahu C., 2005. Bacterial loadassociated with culture of ornamental fishes in glass aquarium. *Environment & Ecology*, **23(1):** 118-123.

Sinha A., 2006. Ornamental fish biodiversity of India. *In* Aquatic biodiversity in India. Published by Narendra Publishing House, Delhi. 213-226.

Ghosh S, Sinha A. and Sahu C., 2007. Dietary probiotic supplementation in growth and health of live-bearing ornamental fishes. *Aquaculture Nutrition*, **14(4)**, 289-299.

Goswami U.C., Basistha S.K., Bora D., Shyamkumar K., SaikiaB.andChangsan K., 2012. Fish diversity of North East India, inclusive of the Himalayan and Indo Burma biodiversity hotspots zones: A checklist on their taxonomic status, economic importance, geographical distribution, present status and prevailing threats. International Journal of Biodiversity and Conservation Vol. 4(15), pp. 592-613.

Moyle, P. B. and Leidy, R. A., 1992. Loss of biodiversity in aquatic ecosystems: evidence from fish faunas. In Conservation Biology: The Theory and Practice of Nature Conservation, Preservation and Management (eds Fiedler, P. L. and Jain, S. K.), New York, Chapman and Hall, pp. 127–169.

Singh, A. K., Ansari, A., Srivastava, S. C., Verma, P. and Pathak, A. K., 2014. Impacts of invasive fishes on fishery dynamics of the Yamuna River, India. Agricul. Sci., **5:** 813–821.

Sinha A. and Asimi O.A, 2007. China rose (*Hibiscus rosasinensis*) petals: a potent natural carotenoid source for goldfish (*Carassius auratus* L.). *Aquaculture Research*, **18**: 1123-1128.

Sinha A., 2008. Ornamental fish of India. Published by Director, CIFE, Mumbai, 254p.

Sinha A., 2009. Breeding of Angel fish. Technical Bulletin, Published by Officer-In-Charge, CIFE, Kolkata Centre.

Sinha A., Prabhakar S.K., and Singh K.K., 2013. Application of thermocol as nest material for breeding of Blue Gourami, *Trichogastertrichopterus*. *J. Indian Fish. Assoc*., **39**:75-80

Sinha A. and Santra S., 2016. Integration of High Priced Small Indigenous Fish with Conventional Carp Culture for Nutritional Security and Rural Livelihood. *International Journal of Agriculture Innovations and Research*, **4**: 960-963

Sinha A., 2017. Evolution, trend and status of ornamental fisheries in India and their commercialization. *In* Social Entrepreneurship in Aquaculture, Editors: V.R.P. Sinha, Gopal Krishna, P. Keshavanath and N.R. Kumar. Published by Narendra Publishing House, Delhi, India, 225-240.

Swain, S.K. and Sahoo, P.K., 2003. Effects of feeding Triiodothyronine on growth, food conversion and disease resistance of goldfish, *Carassius auratus* (Linn.). *Asian Fisheries Science* **16**: 291-298.

Telechea F. and Pascal F., 2014. Levels of domestication in fish: implications for the sustainable future of aquaculture. *Fish and Fisheries*, **15(2)**: 181-358.

Vijula, K. and Sinha, A., 2004. Effect of a probiotic *Lactobacillus* as nutritional supplement on gold fish, *Carassius auratus*. Contributed abstract, *In:* National Seminar on Prospects of Ornamental Fish Breeding and Culture in Eastern and North-Eastern India, 16-17 February, 2004. pp.

Vinod, K., Mahapatra, B.K. and Mandal, B.K., 2003. *Brachydanio rerio* (Hamilton) and *Danio dangila*(Hamilton) – Promising species for ornamental fisheries in Meghalaya. Proc. Participatory Approach for Fish Biodiversity Conservation in North East India (Eds. P.C.Mahanta and L.K. Tyagi), National Bureau of Fish Genetic Resources, Lucknow, pp. 225-230.

Walstad, D. L., 2003. Ecology of the Planted Aquarium: A Practical Manual and Scientific Treatise for the Home Aquarist. (2nd edition). Chapel Hill, NC: Echinodorus Publishing.

Watanabe, T.C. and Kron, 1994. Prospects in larval fish dietic. *Aquaculture***124**: 223-251.

Weigel, W., 1973. Aquarium Decorating and Planning. Transl. by G. Vevers. (T.F.H. edition). Neptune City, NJ: T.F.H. Publications, Inc

Wolf, H. T., 1908. Goldfish Breeds and other Aquarium Fishes: Their Care and Propagation. Philadelphia: Innes and Sons.

www.nabard.org

http://nfdb.gov.in/guidelines.htm

https://mpeda.gov.in/MPEDA/#

9781032599311